D1670164

Decision-Making Tools to Support Innovation

SCIENCES

Chemical Engineering, Field Director – Jean-Claude Charpentier

Innovation and Design in Chemical Engineering
Subject Head – Olivier Potier

Decision-Making Tools to Support Innovation

Guidelines and Case Studies

Manon Enjolras
Daniel Galvez
Mauricio Camargo

WILEY

First published 2023 in Great Britain and the United States by ISTE Ltd and John Wiley & Sons, Inc.

ISTE Ltd
27-37 St George's Road
London SW19 4EU
UK

www.iste.co.uk

John Wiley & Sons, Inc.
111 River Street
Hoboken, NJ 07030
USA

www.wiley.com

Library of Congress Control Number: 2022944834

British Library Cataloguing-in-Publication Data
A CIP record for this book is available from the British Library
ISBN 978-1-78945-089-7

ERC code:
PE8 Products and Processes Engineering
PE8_2 Chemical engineering, technical chemistry
SH1 Individuals, Markets and Organisations
SH1_11 Technological change, innovation, research & development
SH4 The Human Mind and Its Complexity
SH4_7 Reasoning, decision-making; intelligence

Contents

Foreword

Christian FONTEIX
ERPI, ENSGSI, Université de Lorraine, Nancy, France

As a specialist in modeling in process engineering (formerly known as chemical engineering), I would frequently address the issue of the use of models in industrial process optimization for the purpose of making a decision in innovation. Usually, this would be a question of optimizing a single function (mono-objective optimization), often cost. At the end of the 1980s, one of my PhD students was developing a diploid genetic algorithm, where genes were encoded in real numbers because this type of optimizer was able to minimize or maximize a continuous or non-continuous function, whose variables could be real numbers or integers. The student told me about a discussion she had with another student at a convention about multi-objective optimization, which was to replace single-objective optimization in the future, and I realized that this replacement was possible with the help of the algorithm that had been developed and the Pareto domination principle. Since then, several of my students have conducted research on multi-criteria optimization (a term preferable to multi-objective optimization) and its industrial applications.

In the late 1980s, I also met Professor Laszlo Nandor Kiss, at the University of Laval, Quebec, Canada, who introduced me to multi-criteria analysis, referred to as decision-making engineering by engineers. I then quickly understood that this tool was complementary to multi-criteria optimization and essential for making decisions in innovation at the industrial level. In fact, for industry leaders, multi-criteria optimization makes it possible to define a set of compromises representing objective information: it effectively serves as a way to reduce the number of alternatives considered to make an innovation decision. But this method will not be sufficient if there is a high number of criteria and alternatives. Decision makers will need to take into account the subjectivity of their preferences. This leads the decision-making process used by industrialists to be modeled by "measuring" such preferences. However, choosing a decision model is not easy because, for example, it is necessary to know whether having a sufficient value for one criterion can or cannot compensate for a low value of another criterion.

Once the choice has been made, it is natural to ask: was this the right choice, the one that best represents that of the decision maker? This means that it is necessary to assess the robustness of the decision in relation to the choice of the model, the values assigned to the parameters defined by the preferences and the uncertainty in the evaluation of the criteria for each alternative. This is not easy to do in innovation decision-making, especially in the current fluctuating industrial context.

In the early 2000s, I met Mauricio Camargo, who was interested in these issues related to decision-making in innovation. Today, he teaches them and conducts research activities in the field. He is a decision specialist in business innovation who has been able to perfectly adapt and change the tools for improving this innovation in an industrial context. On the basis of his research work, in collaboration with other researchers, he now presents the knowledge that is essential for innovation managers in an industrial setting in this book for the ISTE Science Encyclopedia.

It is not a course, strictly speaking, but rather a remarkable manual outlining good practices that are very useful to practitioners. The book clearly highlights how multi-criteria analysis can contribute to the search for compromises when it comes to innovating in various industrial and manufacturing sectors. Thus, this book should be of interest to many readers, as it contains a large number of innovation decision models and preference models.

As specialists in the field, the authors have chosen to present six of these that are among the most frequently used today. On this basis, the book is structured in six chapters, each of which is devoted to a particular technique, which is described, detailed, explained and applied to various problems related to decision-making in innovation: PROMETHEE (including Gaia plane) for an idea selection process during a creativity workshop, AHP (including AHP-DEMATEL coupling) for product design including the search for informed decision-making from the first phases of the design process, Rough Sets (including DRSA) for strategic decision-making in the commercialization phase, MAUT for the construction of a portfolio of adapted and compatible business projects, ELECTRE for the management of human resources and the recruitment process used for new staff members and TOPSIS for knowledge management within the company and in the entire value chain, which is the successive set of value-added activities that make it possible to start from raw materials in order to obtain a finished product with targeted use value for a customer.

This book is a must-read, which is highly recommended for decision makers in industrial and service companies to provide them with solutions and an open-mindedness necessary for their operations involving innovation decision-making. And it is also recommended for academics seeking to discover and apply multi-criteria analysis methods in their teaching and research projects and activities.

Introduction

In our daily lives, we all must make decisions of different kinds. These might include where we go to eat lunch, our next trips, the next new car we will buy, our career development, etc. If the choice is to be made between several alternatives and if it is dependent on multiple different factors, then we are faced with a multi-criteria decision problem. Generally, people tend to reduce the limits of the problem by transforming it into a general decision, in other words, a problem that highlights a predominant factor at the time when the decision is made. However, when this decision involves an important issue, reducing its complexity would not be appropriate because simplifying the problem would also entail a loss of information necessary to make the right choice.

In companies, decision-making is also a very common practice, though one that is complex, because it is necessary to reconcile different factors and points of view. For example, in the development of a current project, the manager is faced with several decisions: from the allocation of resources to the choice of suppliers, the mode of transport to be used or even the method of financing. Not to mention that with the current development of the digital world, the data sources used to qualify and compare the alternatives of choices have multiplied.

In addition, in the specific case of an innovation project, the decisions become more complex. Indeed, by its own nature, the innovation process involves the simultaneous involvement of technical, economic and environmental factors associated with a new product, as well as changes in the company's operating modes, the integration of new players or uncertainties related to a new market.

For all these reasons, making the right decision (and making it on time) can mark the difference between success or failure. That is why being able to rely on decision-making tools and methods is extremely useful to companies. In this book, we will present several multi-criteria decision support methods by analyzing how they are applied in the context of practices specifically related to the innovation process.

I.1. The innovation process and decision-making

In recent years, the management of the innovation process has evolved toward becoming standardized in companies. As a result, companies are seeking to organize themselves to make the innovation process more and more systematic and recurrent.

In the 2000s, innovation studies were focused on the internal processes that characterize the innovation potential of a company. Chiesa et al. (1996) published one of the first research papers on models for assessing the ability to innovate. Their model served as a reference for the development of a trend for assessing enterprises' ability to innovate based on good practices. These models have made it possible to diagnose, understand and improve the ability of companies to innovate through transferring knowledge, methodologies and successes of leading organizations in this field.

The standardization of the innovation process within companies took an important step forward with the creation of the international standard ISO 56002 (2019). This standard provides general guidelines for establishing an innovation management system.

The implementation of this system has potential advantages for companies and is based on many different elements (Figure I.1): leadership (5), which reflects the commitment of management in promoting a culture of innovation; planning (6) to establish the path to follow to achieve the company's innovation objectives; support functions (7) grouping the necessary resources to establish, implement, maintain and improve the innovation management system ; operational activities (8) to implement all actions necessary to manage the system; evaluation (9) and improvement (10) to control the evolution of the innovation management system to ensure its success over time; and finally, contextual elements (4) to make the entire system specific to each company.

Using the ISO 56002 (2019) standard as a reference framework in this book, we will more specifically use a model in line with the foundations associated with it: the potential innovation index (PII) proposed by the ERPI research laboratory of the University of Lorraine (Boly et al. 2000, 2014; Galvez et al. 2013). The first works on PII date back to the 2000s. From that point forward, this index has progressively evolved into its current version, supported by academic research, expert opinions, interviews with the industrial sector and cases of application in real-world situations.

The latest version of the PII is based on multi-criteria approaches, grouping the activities and processes of the innovation management system companies within six major practices.

– Generating new ideas: an innovative company needs to strengthen its ability to generate ideas. The use of creativity techniques, the integration of users and the establishment of a monitoring system help to generate a constant and effective generation of ideas.

– Design: in order to innovate, a company must be able to convert ideas into concrete realities by materializing them. Thus, design is a key step for an innovative company. There are various resources to support design, such as technical resources, methodologies or computer tools.

– Strategy: a company that seeks to develop its innovation potential must look to the future and adopt a forward-looking vision. It must define where its value will be found over the short, medium and long term and establish an appropriate action plan. The company must also anticipate and arbitrate the financial aspects and those related to intellectual property.

– Project management: the success of an innovation project depends on the management of the resources involved: financial resources, technical resources and human resources. The management of an effective innovation project must be flexible to properly manage the uncertainty that is inherent in innovation.

– Human resource management: companies that innovate have an employee profile oriented toward risk-taking, the desire for professional development, resilience and a thirst for change. Human resources management must encourage and support these behaviors, in addition to ensuring the acquisition and monitoring of the skills necessary for the proper daily functioning.

– Knowledge management: developing memory capacity is essential in allowing a company to learn from previously completed innovation projects. This learning allows for knowledge to be shared and reused, as well as for continuous improvements in the innovation process.

PII is an assessment tool that allows companies to measure their capability to innovate through a self-diagnosis[1] based on the assessment of their maturity in relation to the six practices presented above. This diagnosis evaluates the innovation potential of companies by identifying their strengths (drivers) and their difficulties (obstacles), thus allowing for the development of an action plan that is able to improve their capability to innovate.

Thus, innovation is based on taking into account different interrelated elements within a management system to be managed. This requires the implementation of good practices. Although innovation has become an increasingly common activity in companies, it must be recognized that it is an issue that is complex. More specifically, an innovation project is defined by three characteristics (Kapsali 2011): its uncertainty, its complexity and its uniqueness. Indeed, the tasks needed to complete the project are subject to inevitable changes, which give rise to continuous evaluations and iterations of the project. In addition, innovation projects continue to involve more and more stakeholders who must communicate and agree on the direction and progress of such projects. Therefore, decision-making should be streamlined as much as possible and should be shared and accepted by all relevant stakeholders.

1 Self-diagnosis available at: www.innovation-way.com/.

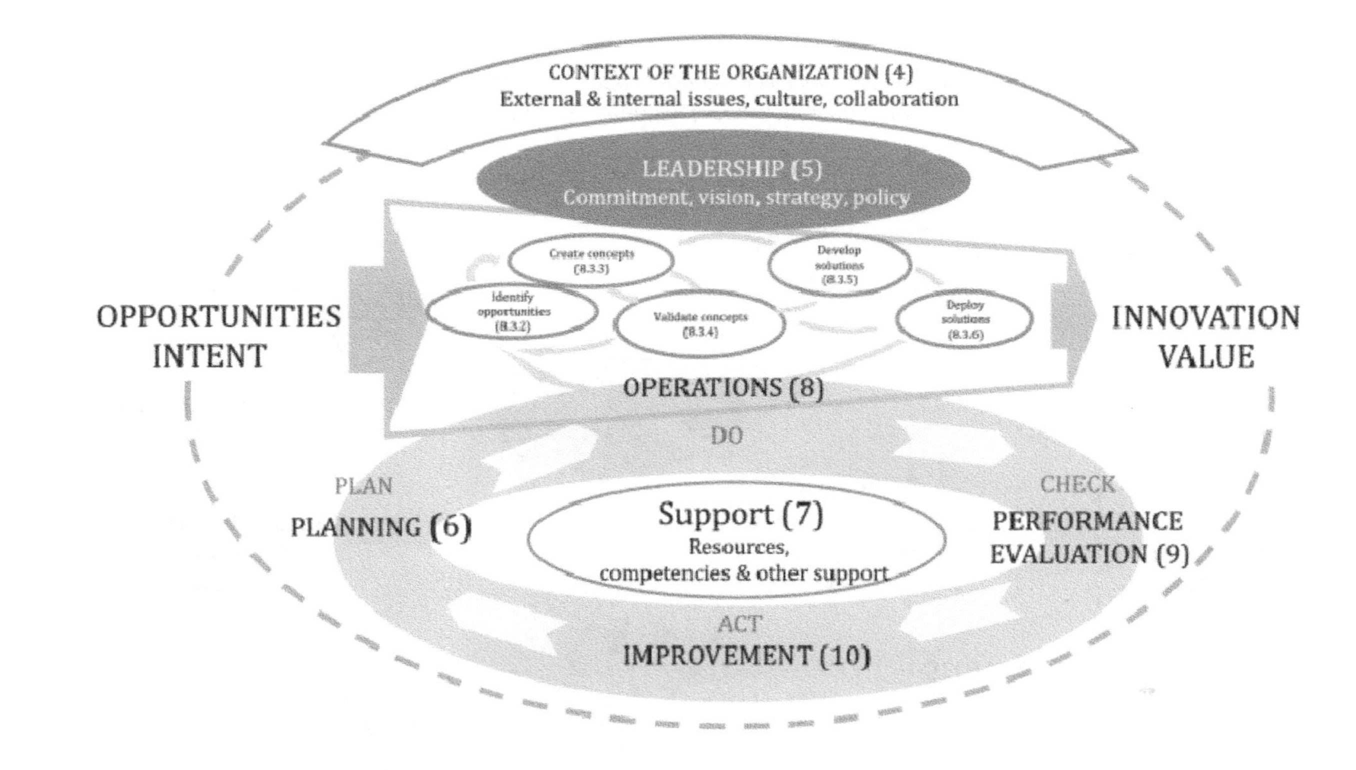

Figure I.1. *Innovation management system – ISO 56002: 2019*

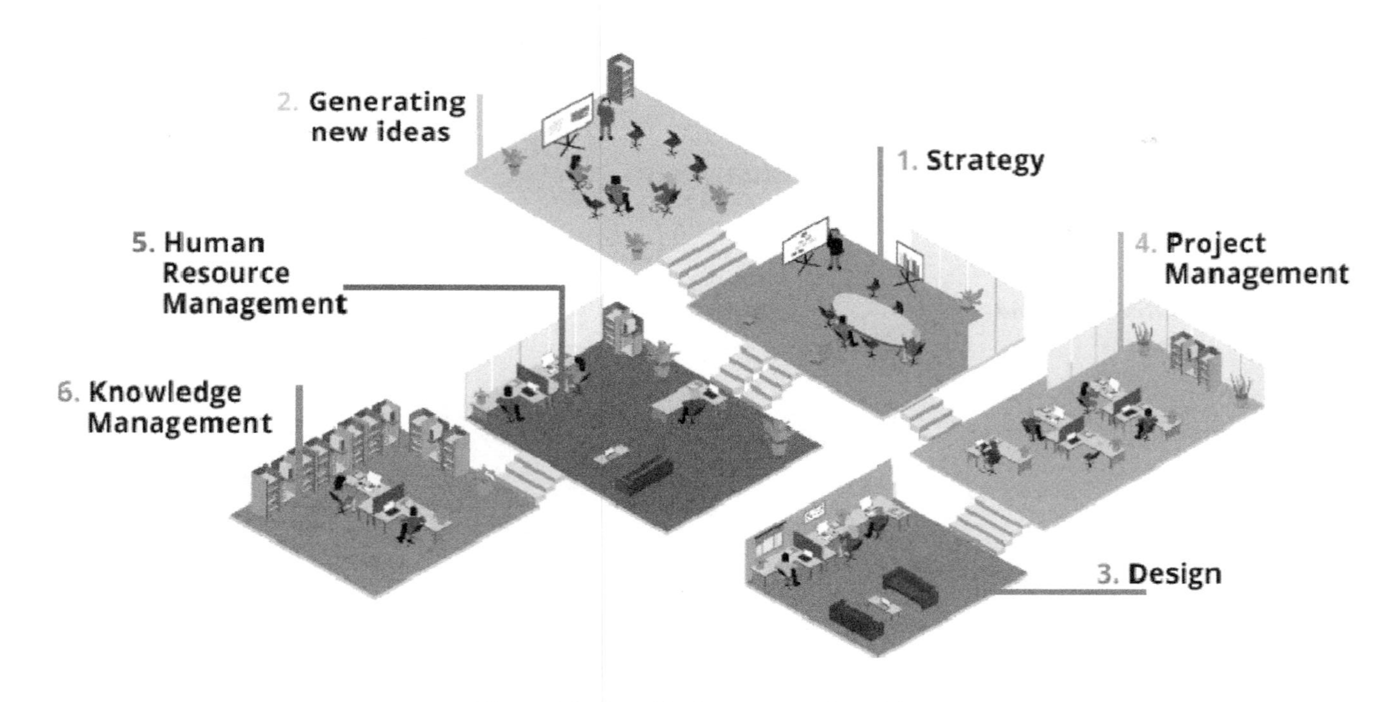

Figure I.2. *Graphical representation of the potential innovation index (source: www.innovation-way.com/)*

Indeed, the innovation process operates in a context of uncertainty, where decision-making depends on several factors. For example, let us suppose that a car company wants to launch a new automotive product line that improves the environmental impact of its vehicles by integrating a solar energy system. The criteria of an environmental nature associated with this decision will be positively impacted, while the economic and technical criteria, for their part, will potentially suffer a negative impact. Making a car with a completely new energy system will likely require a more significant investment, and the technical complexity will also increase. Thus, we will get a car that is environmentally friendly, but is complex to manufacture and more expensive than others. The question then arises: should the company prioritize sustainability, technology or cost? Thus, it is not always easy to find a balance to satisfy several criteria at the same time. It is therefore necessary to make trade-offs and seek a consensus.

I.2. Multi-criteria decision-making

The study of decision-making is multidisciplinary by nature. It has been the subject of study within several disciplines, such as behavioral economics (Thaler and Sunstein 2009) and experimental psychology (Kahneman 2011; Klein 2017). As defined by Gary Klein in his book *Sources of Power: How People Make Decisions*, there are two main strategies that humans use for making decisions. The first is "situational decision-making, or naturalist strategy", in which decision-makers have to make decisions, in real time, under pressure, in dynamic environments, and in which the objectives are not clearly defined (e.g., a firefighter, a member of the military or a doctor in an emergency situation). For this type of decision, the important factors are the experience of the decision-makers, their intuition and their ability to imagine possible scenarios. The second type of decision, called the "rational choice strategy", involves subdividing the problem into criteria in order to structure decision-making. This type of strategy is more suitable when choices need to be justified, when there are conflicts between stakeholders, when the decision-maker is in a decision optimization strategy or when there is a significant computational complexity. In this book, we are interested in the second type of decision-making strategy.

To this end, it is necessary to model the issues involved in the decision, based on criteria to be taken into account, and alternatives to be evaluated. Since the choice among these alternatives is inherently dependent on several criteria, it is necessary to build a model that brings together all these criteria, in order to ensure that they are taken into account when making the final decision. To accomplish this, the notion of multi-criteria analysis is used.

Multi-criteria analysis necessarily leads us to work with several factors at the same time. These factors will allow us to assess all the dimensions that determine the effectiveness of alternatives considered in the decision problem. This group of alternatives can be finite or infinite. If the problem is characterized by an infinite number of alternatives, we are faced with a multi-objective optimization model. In this work, we focus on the study of decision problems characterized by a finite number of alternatives, that is, discrete multi-criteria problems. The objective is to determine the alternative that represents the best compromise to satisfy all the criteria for evaluation. For this, it is necessary to clearly understand the problem and correctly establish the decision to be taken.

An adequate definition of the problem begins with the decision to be made; we need to clarify the question we are trying to answer. Using this question as a starting point, the components of our multi-criteria problem are then defined: the alternatives and the criteria.

The *alternatives* are all valid options that offer a solution to the decision question that is posed. These alternatives must be able to be evaluated by different criteria which will allow us to determine which alternative best meets the preferences of the decision-maker.

The *criterion* must evaluate all the characteristics of the alternatives which are important for the decision-maker. Each criterion will be assigned an evaluation scale to determine whether an alternative is strong or weak regarding this criterion. Each alternative must be evaluated according to all the criteria. If this is not possible, it means that the alternative or criterion has been incorrectly defined. To ensure that the definition of the criteria is correct, we must take into account the following considerations:

– Redundancy (exclusion principle): we must avoid using two criteria defined as different to evaluate the same characteristic. For example, for the prioritization of projects, if we consider the criteria of "costs" and "revenue", we will not need to add a criterion of profitability because it is possible to combine the two previous criteria to arrive at the final one. In this way, we avoid double counting and redundancies.

– Representativeness (principle of completeness): the set of defined criteria should best measure all dimensions that characterize decision-making; ideally, 100% of the problem should be assessed. These first two considerations are combined in the literature within the principle of support for the decision-making process called "MECE". This principle emphasizes the importance of building a benchmark for the criteria that is "mutually exclusive and collectively exhaustive". Next, we will talk about the principle of exclusion and completeness (Lee and Chen 2018).

– Operativeness: all criteria must be measurable by means of an evaluation scale (Roy and Bouyssou 1993). Some criteria are easier to measure than others because, by their nature, it is easier to assign a well-defined value to them. These can include the price of a product (euros), the age of a person (years), the speed of a car (km/h), etc. However, there are other criteria of a qualitative nature that can be difficult to measure, such as the color of a product, the design of a building or the quality of construction work. In this case, it is necessary to define a qualitative assessment scale, which decision-makers will use to assign a value with each alternative for this particular criterion. For example, for the evaluation of the color of a car, if the decision-maker prefers red first, then blue and finally black, we can create a three-level scale and assign the value 3 to red, then 2 to blue and 1 to black. This characteristic refers to the principle of ordinality, that is, if we look at each criterion individually, the alternatives must be able to be ranked from the best to the worst.

– Quantity: we need to define a quantity of criteria that is cognitively manageable for the decision-maker. According to a study by George Miller in 1956, the human brain can work with a maximum of seven elements at the same time (Saaty and Ozdemir 2003); it is therefore recommended to work with a maximum of seven criteria simultaneously. If the problem requires the consideration of more than seven criteria, it is recommended to group them by common axes, for example, a group of economic, technical or environmental criteria. In this way, decision-makers can work by the level of aggregation and divide the problem, thus improving their understanding.

– Temporality: the decision is going to be made during a particular period of time. All the criteria will thus be evaluated simultaneously. In the definition of the problem, it is then necessary to clarify whether the decision will be made in the short, medium or long term.

After defining the set of alternatives and criteria, we need to choose the method of resolution that is most appropriate for the problem (Guitouni and Martel 1998). This choice will depend in particular on the objective associated with the decision-making. The majority of multi-criteria analysis methods have the objective of identifying the best alternative among those considered. Some of the most well-known methods that work in this way are as follows: the weighted average, multi-attribute utility theory (MAUT), analytical hierarchy process (AHP), elimination and choice translating reality (ELECTRE), preference ranking organization method for enrichment evaluations (PROMETHEE) and technique for order of preference by similarity to ideal solution (TOPSIS). By contrast, some other methods are instead oriented toward a characterization of the model by studying the relationships between the criteria (DEMATEL) or by exploiting the information hidden behind the implicit preferences of decision-makers (Rough Sets).

Depending on the purpose they achieve, these multi-criteria methods can be classified into four categories (Wątróbski et al. 2019):

– Description: these methods seek to understand all the ins-and-outs of the decision problem. Descriptive methods provide additional information to assist in decision-making.

– Choice: the expected result is to highlight the alternative that best meets the preferences of the decision-maker. These methods work well if the group of alternatives to be considered is small. They make it possible to differentiate alternatives that intuitively result in a similar preference for the decision-maker.

– Classification or *ranking*: in this case, the result is a list of alternatives ordered according to the degree of preference. This ranking is determined by the overall performance of each alternative considering all the criteria in the calculation of an aggregated score.

– *Sorting*: these methods make it possible to assign the alternatives to previously defined categories. The categories represent profiles of the alternatives behavior in relation to how the decision is made.

Each of these four categories is represented by a variety of methods. But in addition to the objective associated with it, the choice of the specific multi-criteria method to be used also depends on other factors. For example, the robustness of the results may represent a criterion of choice. Depending on the mathematical foundations on which the method is based, the results can be more or less robust, and in the same way, the application of the method can be more complicated depending on the computational steps to be implemented. These mathematical fundamentals also make it possible to consider the compensation phenomenon appearing in certain decision problems. This phenomenon arises when the value of the overall score is significantly influenced by a single criterion. If one criterion is evaluated with a much higher scale than the others, it will cover up and outweigh all other evaluations. Each method is more or less capable of representing and managing this phenomenon of compensation. In the same way, the nature and availability of the input data necessary for the application of the method are to be considered. The graphical data visualization tools associated with the methods are also very useful to improve the decision-maker's understanding. Finally, the availability of a software tool also simplifies the use of a given method by providing a user-friendly interface.

It is not possible to recommend any one method above any other a priori. According to a study by Cinelli et al. (2014), all methods have both strengths and weaknesses. The choice of the method thus depends on the available data, the characteristics of the problem and the results that one wishes to highlight. For example, in the case where a classification or a choice is desired, the AHP method does not need input data because its application makes it possible to construct the weights of the criteria with the decision-maker. This specific step of AHP can also be combined with other methods, thus making it possible to provide a weight vector as input data necessary for their application. On the other hand, in the case where the objective is a description, the Rough Sets method does not use the weight vector in its calculations and therefore does not require input data. With regard to the compensation phenomenon, the ELECTRE, TOPSIS and PROMETHEE methods make this phenomenon more manageable because they operate on the computational principle of partial aggregation (as opposed to total aggregation). Partial aggregation means that initially the alternatives are compared with each other two by two, to identify which one scores higher than the other. From these comparisons, it is possible to aggregate the results to achieve a ranking or choice. In this case, we compare them, then the second step is to aggregate them. In the case of complete aggregation, the alternatives are not compared with each other. Instead, an aggregate score is calculated from all of their evaluations, with these scores then compared a posteriori. In this case, we first aggregate them and then compare them to create the ranking of alternatives.

Finally, it is important to clarify that the multi-criteria methods shown here are constantly evolving. Improvements are regularly made, and new versions are constantly being released. Thus, the objective associated with each method, as well as its computational characteristics, depends on the version used. For example, ELECTRE I and PROMETHEE I have the objective of choosing an alternative, while the ELECTRE II and PROMETHEE II versions that have been proposed as improvements have the objective of the classification of a group of alternatives (see Chapter 5 and Chapter 1, respectively).

	MAUT	AHP	ELECTRE I	PROMETHEE I	DRSA (Rough Sets)	TOPSIS
Objective	Ranking	Ranking	Choice	Choice	Description	Choice
Type of aggregation	Total	Total	Partial	Partial	Not concerned	Partial
Compensation	Yes	Yes	No	No	No	No
Support software proposed	Right choice	Total decision	Decision radar	Smart picker	4Emka	Decision radar
Necessary input data	Utility functions and weight of criteria	None (calculation of criteria weights included)	Weight of criteria	Preferred functions and weight of criteria	None (depending on the case: classification a priori)	Weight of criteria, best and worst alternatives

Table I.1. *Comparison of multi-criteria methods*

In the event that we are not sure which method to use, it is recommended to apply two methods of a different nature in order to compare the results so that the decision made is more robust.

I.3. Multi-criteria methods as a support for the innovation process

In summary, moving from intuitive decision-making to decision-making supported by multi-criteria methods helps in the standardization of innovation processes. Although, in recent years, we have witnessed a development of scientific and pedagogical contributions on the innovation process and on how to manage it, we are still far from achieving a total mastery of the innovation process! In this context, the decision is a key element. Indeed, the innovation process, from its ideation stage to the process of scaling up and industrialization, is composed of a succession of decision-making processes that require technical, economic, organizational, and now sustainable compromises to be made simultaneously. In all innovative activities done by the company (the design of a new product and/or process, digital transformations or the definition of a strategy technology), the stakeholders must seek the best compromise between various and often contradictory dimensions of the same problem.

In this book, the PII and its good practices for innovation as illustrated previously in Figure I.2 will constitute a common thread to illustrate the decisions that can take place around the innovation management system in companies. Each innovation practice that is considered can thus generate decisions whose outcome impacts the company's performance and its capability to innovate.

With this in mind, we have structured this book so that each chapter presents an example of decision-making in connection with one of the PII practices. They all follow an identical structure.

First of all, the context of decision-making is explained so as to better understand the issues involved in the innovation process. Then, the method multi-criteria analysis is described and applied using a case based on a scientific article related to the chosen theme. This step makes it possible to study each method of decision support and to illustrate the principles of operation. The results obtained are then discussed, with a potential opening for greater depth in the analysis of the results or in the application of the method itself. Finally, additional examples of applications are given in the form of instruction manuals, as well as links to open-source software, to help readers progress and guarantee an optional and applied learning of the methods of decision support that are discussed.

The book is structured in the following way:

– Chapter 1 addresses the stage of the selection of ideas in the framework of the practice of Creativity. In this sense, the PROMETHEE method was utilized as part of a creativity challenge;

– Chapter 2 studies the AHP method by applying it to a decision related to the practice of Design: the selection of sustainable processes in the chemical industry;

– Chapter 3 presents a decision illustrating the practice of Strategy by considering the case of the management of the marketing efforts to be made in the cosmetics sector. We study the Rough Sets method;

– Chapter 4 analyzes the decisions involving the management of a project portfolio in the oil industry. The method used is MAUT;

– Chapter 5 focuses on the selection of personnel, a key activity in the practice of managing human resources. This issue will be dealt with using the ELECTRA method;

– Chapter 6 analyzes a decision involving the adoption of knowledge management practices within the supply chain of a manufacturing company. The method of decision support used here is TOPSIS.

Therefore, the goal of this book is to provide readers with a greater familiarity with complex decision-making processes through the use of multi-criteria analysis methods and tools. It thus addresses interrelated research areas, which connect with each other through the development of the stages of the innovation process, the identification of the resulting main decisions and the methods of decision support to be used. Therefore, we will touch on the themes of decision-making and multi-criteria analysis, which are applied to the innovation process in areas such as energy, marketing, sustainable development, logistics.

This book is intended for students in training courses related to engineering sciences, design and strategic management, at the bachelor and masters/engineer level. The book seeks to familiarize them with the simultaneous consideration of multiple criteria during the decision-making process. It is also aimed at corporate practitioners and consultants who are looking to gain expertise.

1.4. References

Boly, V., Morel, L., Renaud, J., Guidat, C. (2000). Innovation in low tech SMBs: Evidence of a necessary constructivist approach. *Technovation*, 20, 161–168.

Boly, V., Morel, L., Assielou, N.G., Camargo, M. (2014). Evaluating innovative processes in French firms: Methodological proposition for firm innovation capacity evaluation. *Res. Policy*, 43, 608–622.

Chiesa, V., Coughlan, P., Voss, C.A. (1996). Development of a technical innovation audit. *J. Prod. Innov. Manag.*, 13, 105–136.

Cinelli, M., Coles, S.R., Kirwan, K. (2014). Analysis of the potentials of multi criteria decision analysis methods to conduct sustainability assessment. *Ecol. Indic.*, 46, 138–148.

Galvez, D., Camargo, M., Rodriguez, J., Morel, L. (2013). PII – Potential innovation index: A tool to benchmark innovation capabilities in international context. *J. Technol. Manag. Innov.*, 8, 36–45.

Guitouni, A. and Martel, J.-M. (1998). Tentative guidelines to help choosing an appropriate MCDA method. *Eur. J. Oper. Res.*, 109, 501–521.

ISO (2019). Innovation management system. Report, ISO.

Kahneman, D. (2011). *Thinking, Fast and Slow*. Farrar, Straus and Giroux, New York.

Kapsali, M. (2011). Systems thinking in innovation project management: A match that works. *Int. J. Proj. Manag.*, 29, 396–407.

Klein, G. (2017). *Sources of Power – How People Make Decisions*. MIT Press, Cambridge.

Lee, C.-Y. and Chen, B.-S. (2018). Mutually-exclusive-and-collectively-exhaustive feature selection scheme. *Appl. Soft Comput.*, 68, 961–971.

Roy, B. and Bouyssou, D. (1993). *Aide multicritère à la décision : méthodes et cas*. Economica, Paris.

Saaty, T.L. and Ozdemir, M.S. (2003). Why the magic number seven plus or minus two? *Math. Comput. Model.*, 38, 233–244.

Thaler, R.H. and Sunstein, C.R. (2009). *Nudge: Improving Decisions about Health, Wealth, and Happiness*. Penguin Books, New York.

Wątróbski, J., Jankowski, J., Ziemba, P., Karczmarczyk, A., Zioło, M. (2019). Generalised framework for multi-criteria method selection. *Omega*, 86, 107–124.

1

The Selection of Ideas During a Creativity Workshop: An Application of PROMETHEE

Innovation arises from the ability to generate and transform an idea into products or services that are then adopted by a market. With this in mind, companies are now implementing more and more initiatives to promote the generation of ideas. Despite the development by companies in recent years of strategies and tools to systematize the innovation process, the success rate of a new product or service from the origin of the idea to the market launch remains very low (at best, one new product is created from out of roughly every 3,000 ideas; Stevens and Burley 1997). The increase in this success rate then depends on several factors: the individual and collective capacity to generate ideas, the relevant evaluation of ideas, the selection of the idea that is best suited to the context, the effective management of the development phase of the project or the realization of an appropriate marketing plan. This means that consistent and transparent decision-making in the initial stages of an innovation project can make the difference between whether that project succeeds or fails. However, making decisions at this stage can be a complex task because many possibilities are still open, while there is little information available to judge them, which makes these first steps very uncertain.

In this chapter, we will focus in particular on the selection of ideas, especially when establishing a creativity workshop. First, we will broadly analyze the challenges involved in the creativity phase. We will then discuss the definition of creativity workshops and the challenges they entail, as well as the main stages for implementing them. Next, we will take a closer look at the process of evaluating ideas, the definition of criteria for evaluating the ideas and finally the choice of the idea to be developed, taking into account the context of the decision.

To illustrate this problem, we will draw inspiration primarily from the works of Gabriel et al. (2017) and Gabriel et al. (2016a), which sought to structure and evaluate these creative workshops. Then, we will explain how a multi-criteria analysis method is applied to select ideas related to new services offered in the ecotourism industry by the city of Leticia in Colombia, within the areas of sustainability. In this particular case, the PROMETHEE multi-criteria method (Preference Ranking Organization Method for Enrichment Evaluations), proposed by Brans in 1982, will be used (Brans and Vincke 1985; Brans and Mareschal 1994). Finally, we will interpret the results obtained for the selection of ideas, and we will discuss other possible applications of PROMETHEE to support decision-making during the innovation process.

1.1. Context and challenges in decision-making

1.1.1. *The phases of a creative workshop*

Nowadays, it is more and more common to organize creativity sessions in all organizations, which may be given various names: "brainstorming, creative challenges, hackathons, so forth". For the most part, these events seek to solve a problem that is framed in a collective and entertaining way, taking advantage of the creative abilities, skills and individual experiences within or outside the organization.

The concept of a creativity workshop is based on the principle of creative problem solving (CPS) introduced by Alex Osborn with the technique of *brainstorming* (Osborn 1963; Sawyer 2012). This technique, and the creative problem-solving approach more generally, established four rules: (1) avoid criticism during the generation and delay it until a later phase, (2) encourage the creation of ideas that are as unusual as possible, (3) prioritize the quantity of ideas since having a large number of ideas increases the likelihood of obtaining quality ideas and (4) bounce around, combine and increase the ideas suggested by others.

Within a company, a creativity workshop consists of bringing together various people (if possible, from different departments and/or various skill backgrounds) to solve a problem by applying creativity techniques. The most widely used techniques are reverse reasoning, analogy, daydreaming or the Scamper technique (Eberle 1972; Van Gundy 1987; De Brabandere 2002; De Bono 2010). The details of the concepts and implementation of these methods are beyond the scope of this book, but interested readers can refer to the references mentioned above.

The purpose of these techniques is to promote discussion, the comparison between different points of view and to reduce latent inhibitions (Carson et al. 2003). Creativity techniques aim to engage people in types of thinking and reasoning that are alternatives to those that are typically used. To this end, the creativity workshop can be broken down into four iterative phases: problem analysis, idea generation, idea evaluation and communication/ implementation (Howard et al. 2008) (Figure 1.1).

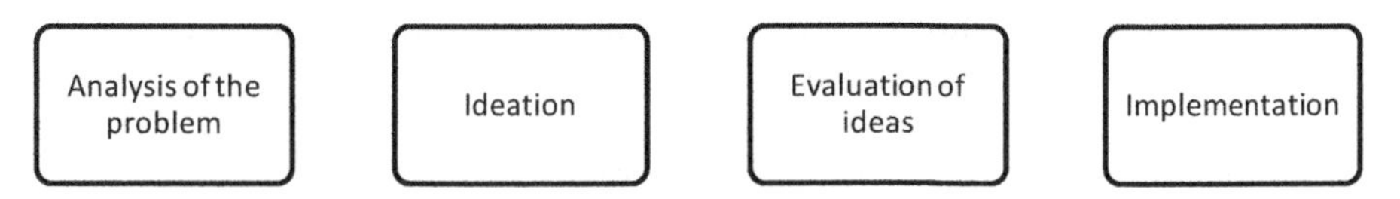

Figure 1.1. *Phases of the creative process*

However, the effectiveness of such events, in terms of actually implemented projects, is still far from satisfactory. For this reason, in recent years, research work has been dedicated to improving this efficiency by promoting a more global vision of the entire process. Indeed, the ideation phase, that is, the workshop itself, represents only one of these stages, which must be completed through the evaluation and prioritization of ideas, or the monitoring of how the innovation project is implemented.

In the following, we will be more interested in the process of evaluating the ideas generated during the workshops.

1.1.2. *Evaluation and selection of ideas*

Each creativity workshop is unique and fits within a specific context, with their sponsors having particular requirements and expectations each time. Without any prompting, the choice of the idea is made instinctively and informally. We want to guide and objectify the selection of ideas based on the information that is available.

Traditionally, after a large number of ideas have been generated as a result of the application of creativity techniques by the participants, one or more ideas are chosen from the mass of ideas produced to be expanded on. This decision-making process involves the formulation of assessments and compromises, as well as the making of decisions. The different approaches that have been identified in the literature to assist this evaluation can be grouped into three main techniques (Westerski 2013): the assessment/evaluation of ideas, computer-assisted processing and the filtering and re-grouping of ideas. The idea assessment, the most common technique, is carried out by an evaluator to enrich the ideas based on the objectives of the organization and its current needs.

Although the evaluation of creative production through a global and subjective assessment by experts can be rich in terms of feedback, it requires a great deal of time to process the information. This processing is all the more difficult given that the definitions of the notions of what can be considered creative are not necessarily shared by all participants involved in this evaluation. Therefore, to systematize the assessment, the hypothesis is to use multi-criteria analysis methods.

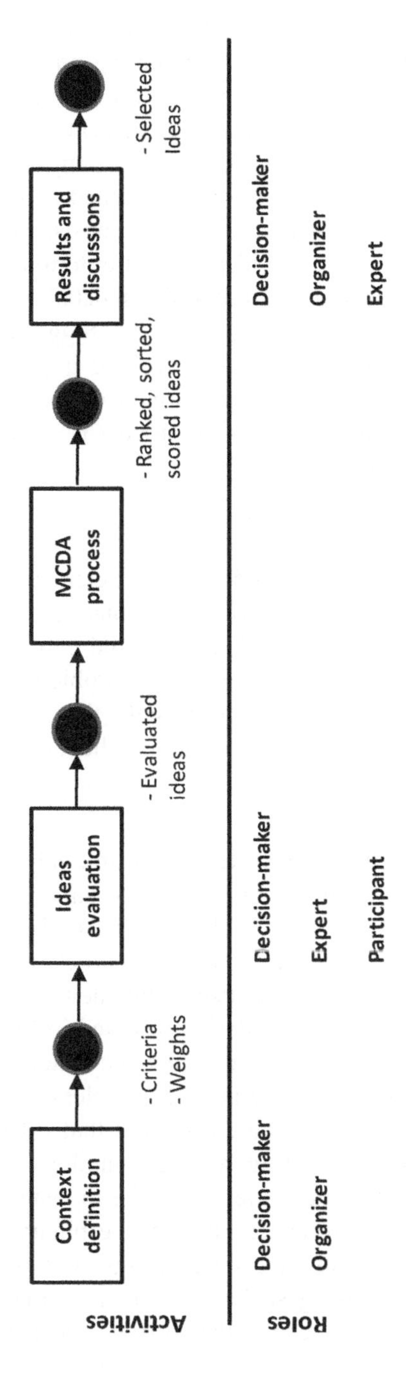

Figure 1.2. *Representation of the idea evaluation approach*

Multi-criteria analysis methods will in effect allow the decision maker to determine the best compromise in terms of combining criteria because there are no single combinations that are suitable for all cases. In addition, the process of formatting the problem provides essential information to understand the evaluation and frame the decision-making process (Nemery et al. 2012). The evaluation approach can be represented as a sub-process of that of the creativity workshop (Figure 1.2). It consists of four stages: definition of the context, evaluation of ideas, multi-criterion processing and discussion. The following subsections will describe in detail the nature of each of the steps in the process of evaluating the ideas suggested above. It is important to emphasize that the participants involved in each stage of the process assume various roles: decision maker, organizer, expert, etc. This can in fact have implications for how the decision process is carried out, since it can be carried out individually or collectively.

1.1.2.1. *Definition of the context*

The context of a problem is made up of different elements, including the organization's business sector, the skill sets that are available, the organization's strategy and the company's culture and expertise in terms of innovation management. The formalization of the problem, its constraints, and its needs makes it possible to correctly state the problem and thus determine the different criteria necessary to evaluate the ideas.

1.1.2.2. *Evaluation of ideas*

The multi-attribute (or multi-criteria) definition of creativity holds that a product must be new, and it must also possess another attribute that guarantees quality (Kudrowitz and Wallace 2013). In order for an organization to make a real contribution to the market, an idea must be applicable, useful and feasible – while also being new (Acar and Runco 2012) – which leads to the evaluation criteria presented in Figure 1.3. Thus, we may assume that decision makers can select the criteria to be applied from a list, or define their own criteria. Nevertheless, when selecting/defining evaluation criteria, it should be considered that a large number of criteria are not suitable for evaluating a large number of ideas (Riedl et al. 2010). From a pragmatic point of view, in order to evaluate the criteria, the evaluator needs an evaluation scale. This can be binary, resulting in a classification by aggregating votes or giving ratings according to a Likert scale. It should be noted that using scales that are too small and a single evaluation parameter can lead to results that differ little from selecting at random (Riedl et al. 2010).

A study on the criteria used by a company to evaluate ideas as part of an innovation approach offered seven criteria: (1) the alignment of ideas with the company's strategy, (2) the feasibility of the idea, (3) the return on investment of the idea, (4) the environmental impact, (5) the social impact, (6) the improvement of parallel projects, (7) other tangible impacts (Correa and Danilevicz 2015).

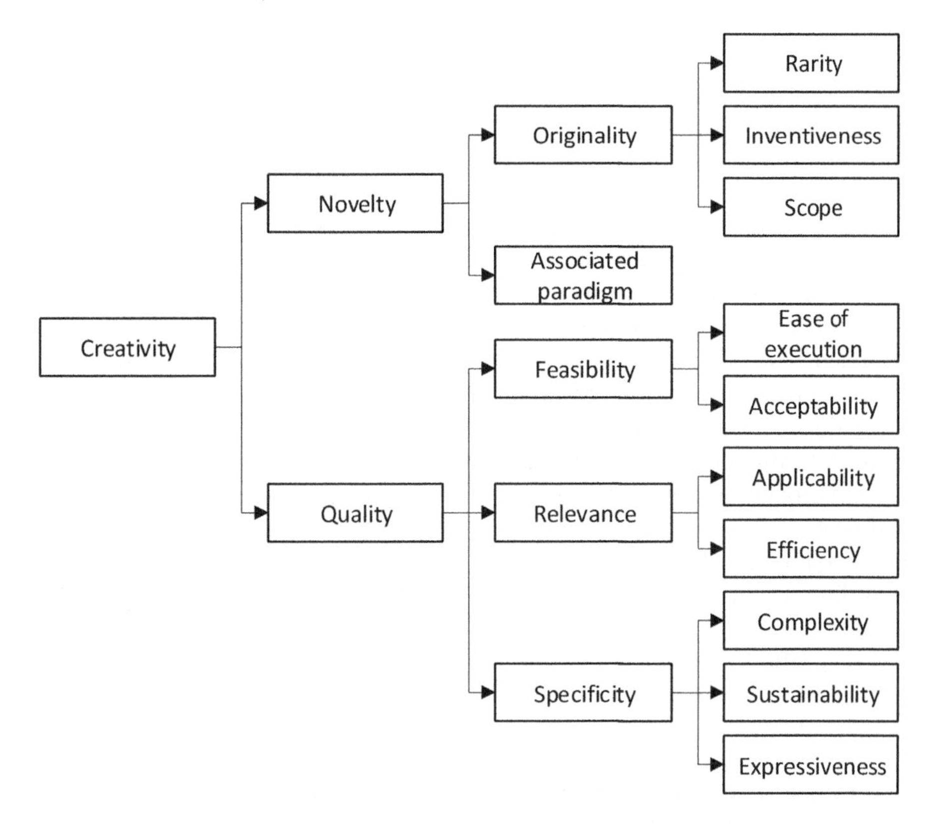

Figure 1.3. *Synthesis of the adapted evaluation criteria from Verhaegen et al. (2013)*

Once the set of criteria has been chosen, they must then be weighted relative to each other. Depending on the objectives of the creativity workshop and the opinion of decision makers, the criteria are not all of the same importance. To determine the weighting of the criteria, two solutions may be used: the decision maker explicitly defines the weight of the criterion, or the decision-making scheme is obtained by applying multi-criteria analysis methods to a series of ideas. An explicit definition is done by assigning a weight between 0 and 100%, with the sum of the criteria being made equal to 100%. For the indirect definition mode, different methods exist to determine the decision-making scheme to be used by decision makers, and the weightings of the criteria that they apply intuitively (see section 2.4).

1.1.2.3. *Multi-criteria analysis*

In order to better understand how the evaluation of ideas is made, three evaluation scenarios are proposed: a single evaluator, an evaluation by aggregation of points of view or a consensus.

– The simplest scenario is that of the single evaluator, which involves either the decision maker or an expert. This scenario can be implemented during the generation of ideas during the creativity workshop or once the set of ideas has been generated. In the case of a prolific creativity workshop, this approach results in the evaluation of an important number of ideas, and is therefore time-consuming. In addition, despite the formalized criteria, the evaluation of ideas from a single point of view can be quite critical. A solution to this problem would be for the ideas to be pre-selected, either by the participants or automatically. The involvement of the participants in the evaluation could be done during activities dedicated during the idea generation phase. Depending on their qualification, they could classify ideas according to groups, map the space of ideas using "mind-mapping" or evaluate ideas using criteria that are specifically dedicated to them. This approach allows for an initial sorting of ideas, to facilitate the task of the evaluator. In a certain way, the participants also become evaluators in a group evaluation configuration.

– In the case of an evaluation carried out by several people, the aggregation of the evaluations can be done in different ways, depending on these people's roles and the expertise of the evaluators. Several evaluators will individually evaluate the same criteria or criteria specifically adapted to their expertise. In this case, the points of view will be varied, unlike the approach with a single evaluator. If these evaluators evaluate the same ideas, the question of which ideas to evaluate first remains relevant. The solution where workshop participants, "ideators", are involved can also be applied. In order for the evaluation of the set of ideas to proceed more quickly, evaluators can evaluate different ideas, which is equivalent to increasing the performance of the single-evaluator approach.

– The consensus scenario consists of collectively meeting the criteria for each of the evaluated ideas. The evaluators discuss the idea based on each of the criteria. This can generate paths for improvement, but is also very time-consuming. This method introduces biases due to the capabilities for negotiation and speaking specific to each individual. This scenario is only applicable to a small number of ideas.

1.1.2.4. *Results and discussion*

Once the ideas have been evaluated, the data generated by the evaluation are processed using a multi-criteria analysis method. This calculation step leads to either sorting, ranking, or a classification, depending on the multi-criteria analysis method used. If digital tools are used to carry out the calculations, this step is exclusively computer-based, and therefore virtually invisible in terms of time. The multi-criteria analysis method applied depends on the number of evaluators (one or more evaluators per idea), the weight of the criteria, the nature of the desired result and potentially the decision-making model (preference functions) of the decision maker.

1.2. The PROMETHEE method

The PROMETHEE method belongs to the family of over-ranking methods. The method was proposed by Brans in 1982. PROMETHEE, as well as other upgrading methods, is based on the principle of pairwise comparisons.

The proposal as made by its authors seeks for the PROMETHEE method to be more intuitive, since decision makers naturally tend to compare each action broadly with another in order to determine the best one, and not individually, criterion by criterion.

The core concept of this method is the notion of the preference of the decision makers. Thus, the degree of preference is determined as a function of the difference between two alternatives with regard to a given criterion. Thus, the PROMETHEE method allows the decision maker to compare and aggregate preferences between alternatives from the set of criteria, regardless of the units or scales of the criteria. As we will see later, once they are evaluated and aggregated, these preferences make it possible to systematically compare actions with each other and to establish rankings, while avoiding the disadvantages of other methods, such as compensation and incomparability.

The compensation essentially lies in the fact that for a given alternative, a very low performance on one criterion can be completely compensated for (i.e. eliminated) by a good performance on one or more of the other criteria. This is the case with weighted averages, or other methods of total aggregation (such as MAUT or OWA). Although compensations may not be a major disadvantage for some decision-making problems, in some cases they are not acceptable. For example, in evaluating a product using sustainability criteria, the environmental impact or toxicity of a product cannot be offset by a cost that is by far more favorable than this of the competitors.

For these reasons, in addition to its easiness to be implemented, PROMETHEE is one of the most widely used and applied methods.

1.2.1. *Methodological concept: the preference function*

As described earlier, the PROMETHEE method is based on the comparison between alternatives with regard to a set of criteria. This comparison will allow us to determine the difference between these two alternatives on each criterion. However, the difference alone is not enough to ensure that one alternative will be preferred over another. Indeed, the value of this gap has a meaning for the preference of decision makers. That is to say, for low, intermediate, or high deviation values, the degree of preference may change, depending on each decision maker.

Thus, the preference function reflects the difference between the evaluations or performances of alternatives for a given criterion, based on a degree of preference (see Figure 1.4). This degree of preference is represented by an increasing function of the deviation: the smallest deviations will lead to lower degrees of preference, and the largest deviations to stronger degrees of preference. An exception to this is the case of an absolute preference (see Figure 1.5) in which the preference is constant for any deviation value.

The degree of preference is expressed on a normalized scale between 0 and 1 (or, preferably, as a percentage). Figure 1.4 shows an example of a preference function for a given criterion. Each criterion used in the decision-making model should have its own preference function, created together with the decision maker.

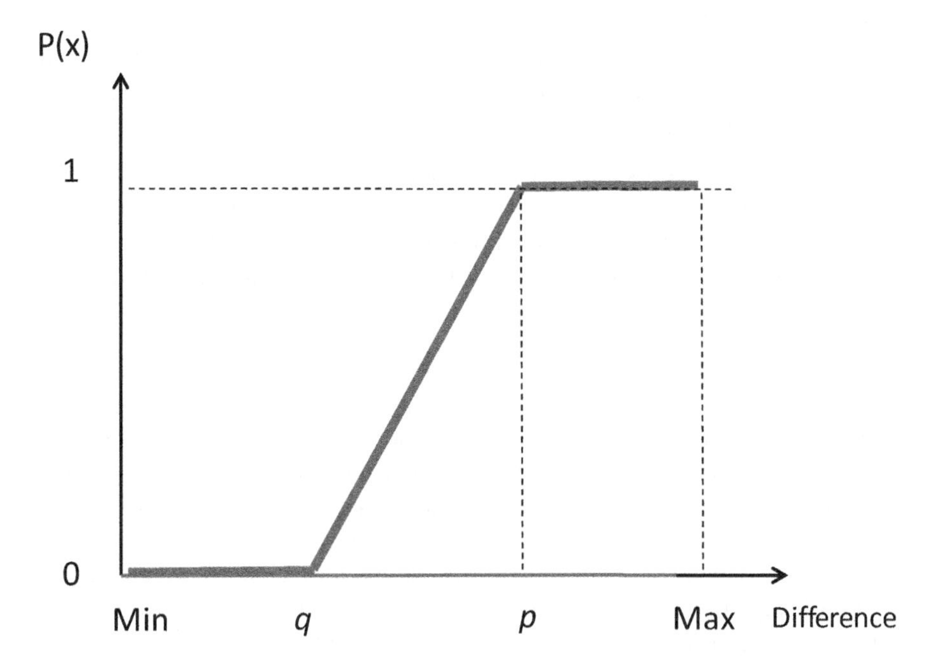

Figure 1.4. *A representation of the preference function for a given criterion*

In this way, in Figure 1.4:

– the Y axis shows the degree of preference of the decision maker $P(x)$. Its value can range from a preference that is null, at zero (0), up to a full preference, at one (1);

– the X axis shows the evaluation differences between the two alternatives considered. This value varies between the minimum deviation value (Min) (which can be zero (0) in the case of a tie between the two alternatives), and up to a maximum deviation value (Max), when we compare the alternative with the lowest value with the alternative with the highest value, on the same criterion. This means that for negative deviation values, the preference value is 0 (the indifference zone).

Also, we can identify two parameters associated with the preference function:

– the threshold of indifference (q) is the largest value considered as negligible by the decision maker when comparing the performance of two alternatives for the same criterion. Thus, all deviations between two alternatives less than q are considered insignificant by decision makers. It can then be concluded that the two alternatives considered have an equivalent performance on this criterion, according to the decision maker;

– the preference threshold (p) is the smallest value considered decisive when comparing the performance of two alternatives for the same criterion. That is, for values beginning at this threshold, decision makers are certain to prefer the alternative that has the best performance (at total preference; $P(x) = 1$).

The difference between the preference limit (p) and the indifference limit (q) will generate an intermediate preference zone whose values will be calculated through the mathematical function associated with the shape of the curve. For example, in Figure 1.4, the intermediate area is a straight line; therefore, we can conclude that the associated mathematical function will be linear: $y = ax + b$. In addition, we know two points: (q, 0) and (p, 1). With these data, it is possible to calculate the parameters a and b by creating a system of equations. Thus, if we let e be the difference between two alternatives, the preference function will be determined by:

$$P(x) = \begin{cases} 0, \ldots \text{ if } e \leq q \\ (e - q/p - q), \ldots \text{ if } q < e < p. \\ 1, \ldots \text{ if } e \geq p \end{cases} \qquad [1.1]$$

The method proposes a set of preference functions as shown in Figure 1.5. The choice of a preference function must reflect the best possible representation of decision makers' preferences.

Thus, as shown in Figure 1.5, the first case shows an absolute preference (a). In this case, regardless of whether the gap is large or small, the decision maker will always fully prefer the alternative that is best evaluated in this criterion. The second case (b) is a linear increasing preference function, that is, the preference increases in proportion to the difference. Next, the V-shaped preference function (c) has three zones, which are determined by equation [1.1]. In the U-shaped preference function, the boundary between indifference and absolute preference is equal ($p = q$), where all deviations less than this limit represent a null preference of zero, and all the higher deviations show a total preference of 1. The fifth is the stepwise function (e). This function has three zones, where if the deviation is less than or equal to the limit q, the preference is zero, while if the deviation is between the limit q and the limit p, the preference will be an average of 0.5; and if the deviation is greater than the limit p, the preference will be equal to 1. Finally, (f) shows an increasing Gaussian function, but it has no threshold of indifference or defined preference.

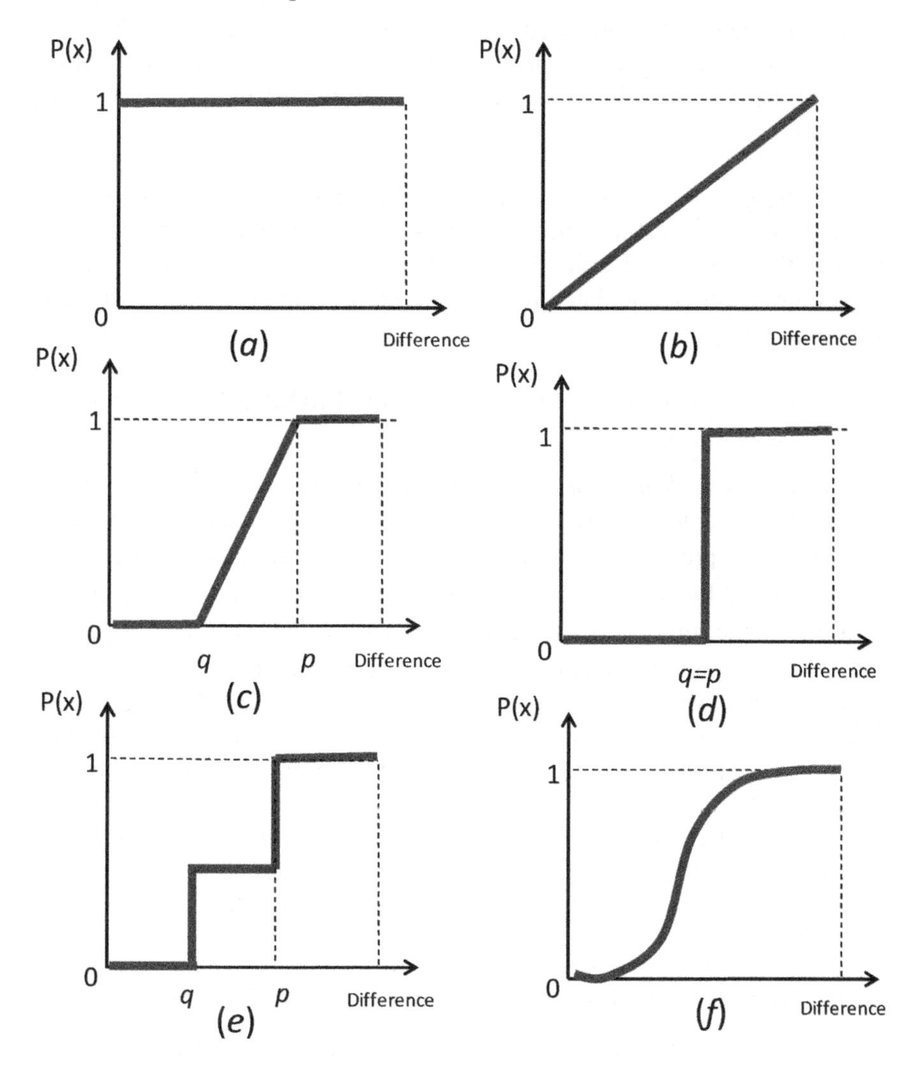

Figure 1.5. *Types of preference functions: (a) absolute; (b) linear; (c) V-shaped; (d) U-shaped; (e) plateau; (f) Gaussian*

1.2.2. *Application process*

As with most decision support methods, $A = \{a_{1,} \dots, a_m\}$ is the group of alternatives m that must be classified, and $F = \{f_{1,} \dots, f_n\}$ is the set of n criteria that need to be optimized. The situation of the decision to be made can be summarized by a decision matrix $(m \times n)$ where the elements of the matrix are the values of the evaluation $f_j(a_i)$ for the alternative a_i, according to criterion f_j.

Thus, the generation of the ranking of the alternatives first involves an evaluation on the part of the evaluators. Once all the evaluators have examined the different criteria, the average of the scores for each alternative and each criterion is calculated to generate the average evaluation value, in other terms $f_k(a_j)$.

From this information, the application of the PROMETHEE method can be described as follows:

– *Calculation of degrees of preference for each pair of alternatives and by criterion*. The alternatives are compared two by two for each of the criteria. The degree of preference P_{ij}^k is determined according to a preference function defined by decision makers, which will depend on the difference in the values of the alternatives for a given criterion, that is, $f_k(a_i) - f_k(a_j)$. As noted in the previous section, the degree of preference is expressed by a number falling between the range [0, 1]. The value of 0 represents indifference or non-preference, and the value of 1 means strict preference. Therefore, a preference function must be defined by the decision maker. For example, for this illustration, the preference function applied may consist of preferring the alternative with the highest score, no matter how small the difference in score between the alternatives (absolute function (a) according to Figure 1.5), that is to say $P_{ij}^k = 1 \; si \; f_k(a_i) - f_k(a_j) > 0$.

– *Definition of the weight vector by the decision maker*. This vector measures the relative importance of each of the criteria, $W = \{w_1, \dots, w_n\}$.

– *Calculation of degrees of multi-criteria preference π for all alternatives*:

$$\pi(a_i, a_j) = \sum_{k=1}^{n} w_k P_k \left(f_k(a_i) - f_k(a_j)\right) \text{ with } \pi(a_i, a_j) \in [0; 1]. \qquad [1.2]$$

The multi-criteria degree of preference $\pi(a_i, a_j)$ is the measure of the preference of decision makers for alternative a_i compared with alternative a_j taking into account all the criteria f_k and their weightings w_k.

– *Calculation of flows of preferences*: the functions of preference make it possible to systematically compare the alternatives with each other. Now, we need a way to summarize the results of all these comparisons. For this purpose, we calculate the preference flows. The positive flow (ϕ^+), or outgoing flow, is the degree of preference by which this action is preferred over other actions on average. The higher the flow ϕ^+, the better the alternative:

$$\phi^+(a_i) = \frac{1}{k-1}\sum_{j=1}^{n} \pi(a_i, a_j) \text{ with } \phi^+(a_i) \in [0; 1]. \quad [1.3]$$

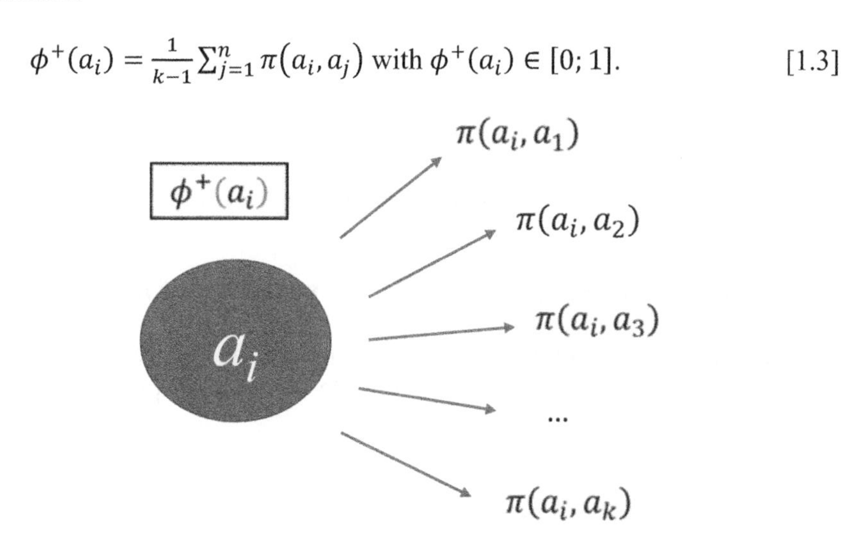

Figure 1.6. *Outgoing flow for alternative a_i*

Negative flows (ϕ^-), or incoming flows, represent the degree of preference by which the other alternatives are preferred to this alternative, on average. The smaller the flow ϕ^-, the better the alternative:

$$\phi^-(a_i) = \frac{1}{k-1}\sum_{j=1}^{n} \pi(a_j, a_i) \text{ with } \phi^-(a_i) \in [0; 1]. \quad [1.4]$$

– The calculation of the net flow, or overall flow, groups the positive flow and the negative flow:

$$\phi(a_i) = \phi^+(a_i) - \phi^-(a_i) \text{ with } \phi(a_i) \in [-1; 1]. \quad [1.5]$$

From the calculation of the overall flow for each alternative, it is then possible to produce a ranking. The best alternatives are those with the highest overall flow.

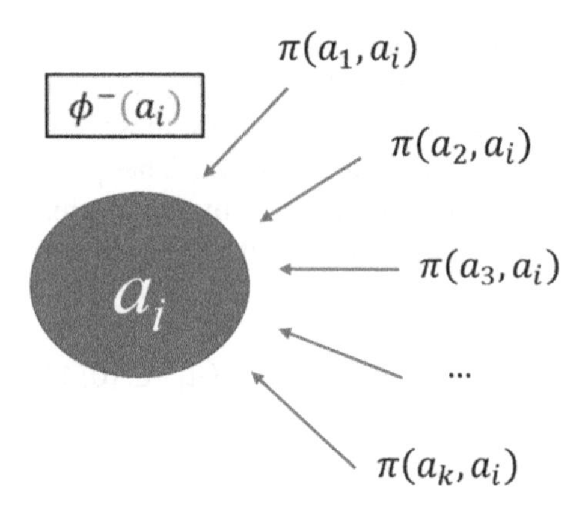

Figure 1.7. *Incoming flow for alternative* a_i

1.3. Application of PROMETHEE to the selection of ideas

In order to illustrate the approach proposed in Figure 1.2, we will present a case study in which we were able to use it in practice (Gabriel et al. 2016a).

1.3.1. *Context of the workshop*

The Amazon region in South America remains one of the most preserved places in the world. However, due to its environmental and cultural diversity, over the past few years the region has become an attractive destination for tourists. During the first decade of the 21st century, there has been a significant increase in the number of tourists visiting the most important cities of the region, such as Leticia (+500%), Iquitos (+200%) and Manaus (+300%) (Obando Lugo et al. 2010). Despite the economic growth that this influx of tourists brings to the region, there is growing concern about the sustainability of this development. This concern more specifically involves the environmental impact, the marginal inclusion of local indigenous populations due to this phenomenon and the cultural impact to such populations through their contact with tourists (Ochoa 2008; Craven 2015). It is in this context that the city of Leticia and its most important stakeholders (the university, local authorities and tourism agencies) seek to explore creative solutions to support the development of eco-tourism while finding the best compromise between local development and the respect for the environment and the traditions of the people.

The topic of the creativity workshop was defined as follows: "exploring new products and services to promote ecological tourism from a sustainable point of view in the Amazon region".

Two workshops were organized in coordination with the National University of Colombia: the first with a group of 35 students in Bogota, and the second with a group of 25 participants (including local players and students) in Leticia. Each workshop began with a half-day meeting, consisting of a short introduction regarding the objectives and framework of this workshop followed by a short team cohesion exercise to create an atmosphere conducive to creativity. Then, two divergence-convergence loops were performed as part of a typical creative dynamic. In this exercise, each of the groups generated ideas, chose one from the generated ones, and presented it to a jury in a brief presentation. The panel was made up of five people, either teachers-researchers or stakeholders in the city. They individually evaluated each of the ideas that were presented, on the basis of common selection criteria.

In parallel with this creativity workshop, the idea evaluation approach presented by Figure 1.2 was also carried out.

1.3.1.1. *Step 1: definition of the set of criteria and scales*

Following the methodology presented above, the first step is to define the evaluation approach. Although no specific expectations were expressed by the stakeholders during these workshops, the definition phase therefore began with the identification of the six selection criteria used by the jury to evaluate the ideas: the originality of the idea, the added value created, the difficulty of implementation, the risks induced by the idea, sustainability, and finally, the quality of the presentation. The evaluation scale has been defined from 0 to 5 for each of the criteria (with 5 being the maximum score):

– originality: it evaluates the novelty, rarity, and non-standard nature of an idea according to the general population of ideas;

– sustainability: it evaluates how the idea will be implemented while minimizing its impact on the environment and indigenous culture, and respecting local living standards;

– added value: it evaluates whether the idea will create value within the local population and the region in general, though not necessarily in terms of monetary units;

– difficulty of implementation: evaluates the level of complexity to implement the idea with local resources, and whether it does not induce constraints, or does not violate known regulations or standards;

– risks: these measure the uncertain aspects of the implementation and the acceptability of the product, service, or project resulting from the idea;

– quality of presentation: this refers to the quality of the presentation of the idea carried out by the groups in front of the jury, regardless of its content.

1.3.1.2. *Step 1a: obtaining the weighting of the criteria*

With regard to the weighting of the criteria, the preferences of the stakeholders have not been sufficiently explained. In addition, it was not possible to define a panel of ideas for determining the decision-making scheme. Therefore, all criteria were considered equally important (16.67%). Nevertheless, in order to test the suggested methodology, two additional evaluation scenarios have been added. The first favors originality, while the second centers on the difficulty of implementation. These two different evaluation scenarios have the goal of visualizing the impact of the evaluation strategy on the ranking of the ideas.

1.3.1.3. *Step 2: evaluation of ideas*

Students from different engineering disciplines, as well as non-student participants, were divided into five groups. At the end of the 2 days of work, a total of 88 ideas were produced. From these ideas, each group selected an idea to present to the five judges, who individually evaluated each idea of each of the groups on the basis of the six criteria given earlier.

Five ideas seeking to promote ecological tourism in the Colombian Amazon were selected to be presented to the jury:

– G1–Air Amazon: the concept is to offer an experience of the Amazon as seen from the sky. Sightseeing tours are organized in an environmentally friendly aircraft that operates using solar energy.

– G2–Kia Expedition: an expedition based on a survival game where participants can learn the culture of the peoples of the Amazon. It is a physical, cultural and environmental challenge that simulates the lifestyle of the indigenous inhabitants of the region.

– G–Amazonian Visitor Center: this proposal is for the creation of a training center within the airport, where all tourists entering into the territory will have to pass a certification on the local and regional environment and on good practices for protecting nature. The objective is to practice sustainable tourism.

– G4–Cupid on Amazonia: this would offer an experience specially designed for couples who would like to enjoy activities in the middle of nature, while respecting the environment. The services offered would include, among others, special offerings for a honeymoon.

– G5–"No cash": this solution seeks to promote the bartering economy. Tourists would exchange their products with the inhabitants of the Amazon to obtain typical products from the region.

Criterion		Originality	Added value	Difficulty of implementation	Risks	Sustainability	Presentation
Weighting set 1 (%)		16.7	16.7	16.7	16.7	16.7	16.7
Weighting set 2 (%)		50.0	10.0	10.0	10.0	10.0	10.0
Weighting set 3 (%)		10.0	10.0	50.0	10.0	10.0	10.0
Name of idea	(G1) Air Amazon	3.3	3.1	2.9	3.3	3.1	4.1
	(G2) Kia Expedition	4.1	4.1	3.4	3.7	3.7	4.9
	(G3) Amazon Visitor Center	4.3	3.9	3.3	3.6	3.6	4.1
	(G4) Cupid on Amazonia	4.1	3.4	2.9	2.7	3.1	4.0
	(G5) "No cash"	4.7	3.4	3.9	3.3	2.9	3.7

Table 1.1. *Details of the evaluation results for the ideas panel*

In order to illustrate the dynamic nature of the method applied on the basis of the evaluation strategy, three different sets of weightings were tested to evaluate the ideas. The first, which was the product of a context of the creativity workshop with little formalization, implies that the criteria all have the same importance for decision makers, that is, they each have a weight of 16.67% each. The second and third, used only to test the methodology, consider both the originality and difficulty of implementation at 50% each. The assigning of the weightings according to the evaluation strategies, as well as the averages of the juries' scores for the five ideas, are presented in Table 1.1.

1.3.1.4. *Step 3: processing through multi-criteria analysis*

Once the five alternatives had been evaluated by the jury using the set of criteria, the method for applying the PROMETHEE method as described in section 1.2.1 was used.

	Weighting set 1		Weighting set 2		Weighting set 3	
	Net flow ϕ	Ranking	Net flow ϕ	Ranking	Net flow ϕ	Ranking
G1	−0.383	5.0	−0.590	5.0	−0.42	5.0
G2	0.445	1.0	0.267	2.0	0.327	1.0
G3	0.225	2.0	0.235	3.0	0.145	3.0
G4	−0.316	4.0	−0.190	4.0	−0.380	4.0
G5	0.029	3.0	0.277	1.0	0.327	1.0

Table 1.2. *Net flows and ranking of each idea for each of the scenarios*

Table 1.2 shows the net flow nets (equation [1.4]) and the ranking for each idea for the three evaluation scenarios (different weightings of the criteria). Figure 1.8 illustrates the ranking of ideas for evaluation strategy 1. Finally, Figure 1.9 shows the contribution of each of the criteria to the net flow of ideas.

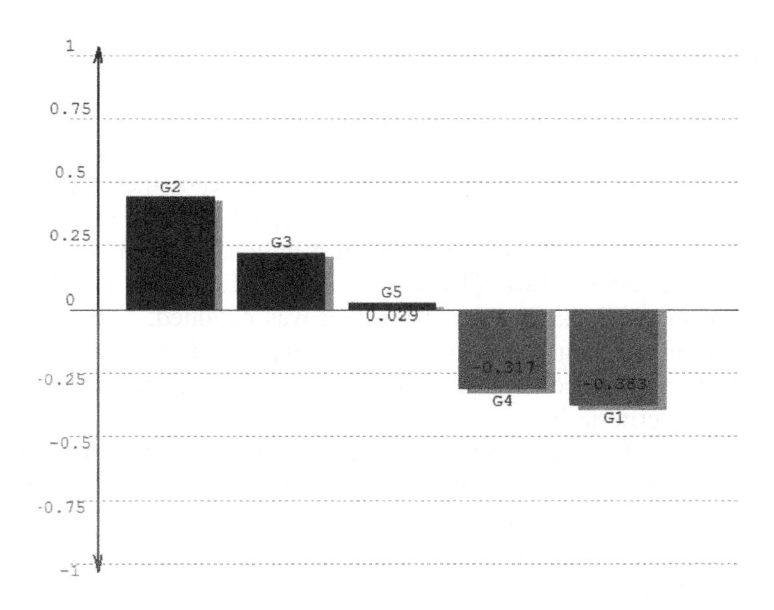

Figure 1.8. *Ranking of the five ideas for evaluation strategy 1*

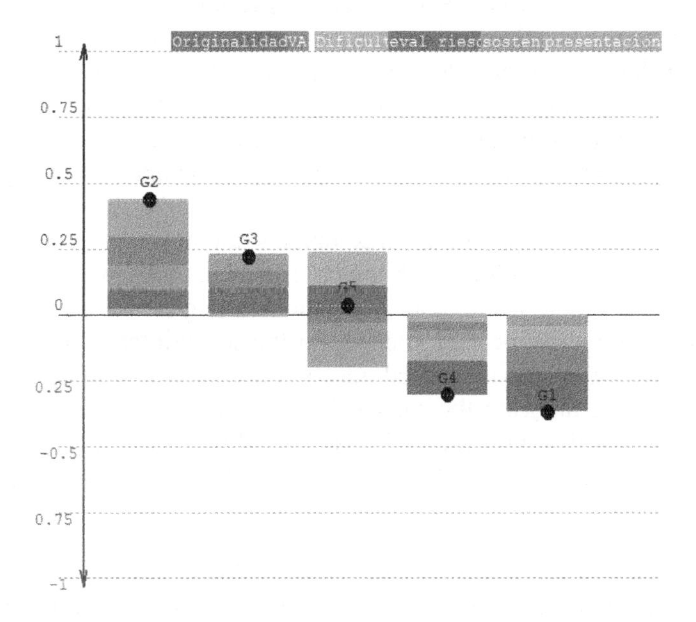

Figure 1.9. *Ranking of the five ideas with the details of the contributions of each criterion for the evaluation strategy 1. For a color version of this figure, see www.iste.co.uk/enjolras/decisionmaking.zip*

1.3.2. *Discussion of the results*

The application of the PROMETHEE method allowed for several different observations to be made. First of all, for scenario 1, in terms of the evaluation strategy (equal-weighted criteria), it appears that the idea of Group 2 (G2 – Expedition Kia) was the highest ranked (Figure 1.8). Therefore, according to the jury, it is the best option on all the criteria and receives the highest net flow (Table 1.2). Figure 1.9 gives an additional indication of the reasons for the ranking that was obtained, as it highlights the contribution of each criterion in the calculation of the flow that was performed. It can thus be seen that for the idea G2, which received the top ranking, all the criteria have a positive contribution to the value of the net flow (the results of its histogram are all shown on the positive axis of the ordinates). In particular, it would appear that the strongest positive contribution is that of the criterion "difficulty of implementation" (shown in green). Thus, it seems that the main asset of this idea would be its ease of implementation, in view of the other ideas, which would perhaps be more complex to implement. If we then look at the bottom of the ranking, the idea with the lowest rating and the lowest net flow is idea G1 (Air Amazon). From Figure 1.9, it can be seen that each criterion has a mostly negative contribution to the net flow value (the histogram is on the negative *Y* axis), in particular the originality criterion, which seems to be its main weak point. Finally, it is interesting to analyze the results of idea G5 ("No cash"), which ranked third. From looking at Figure 1.9, it can be seen that this idea is pulled up by the criteria related to its implementation and originality, while its impact on the environment, and more particularly the quality of its presentation, brings it down. It thus obtains an average-level net flow, according to the scenario for which the criteria are equally weighted. On the other hand, if we look at the other evaluation strategies, we realize that the ranking is changing (Table 1.2). Indeed, for the scenario according to which the originality of the idea is favored, idea G5 comes out on top this time. In the third scenario, favoring the sustainability of the solution, idea G5 ("No cash") once again ties with idea G2 (Kia Expedition). In this case, we find ourselves in a situation where they cannot be compared, with an equivalent net flow.

Thus, the definition of different scenarios highlights the sensitivity of the results obtained according to the chosen evaluation strategies. The weightings assigned to the criteria are therefore of great importance and can provide a very fine level explanation of the preferences of decision makers to better inform their decision.

1.4. To go further

1.4.1. *The Gaia plane*

An additional tool within the PROMETHEE method is a representation on the Gaia plane (Geometric Analysis for Interactive Aid) which allows for a graphical visualization to better understand the decision (Figure 1.10). This Gaia plane thus makes it possible to simultaneously view both the criteria (vectors) and alternatives (points), and therefore performs a more complete interpretation of the results of the evaluation.

The Gaia analysis is the result of the application of the alternatives of the statistical method of principal component analysis (PCA), popular in multidimensional data analysis, to the net flow nets.

Thus, the Gaia plane represents the first two main components of a multi-dimensional space, which ensures that a maximum amount of information is represented in this plane.

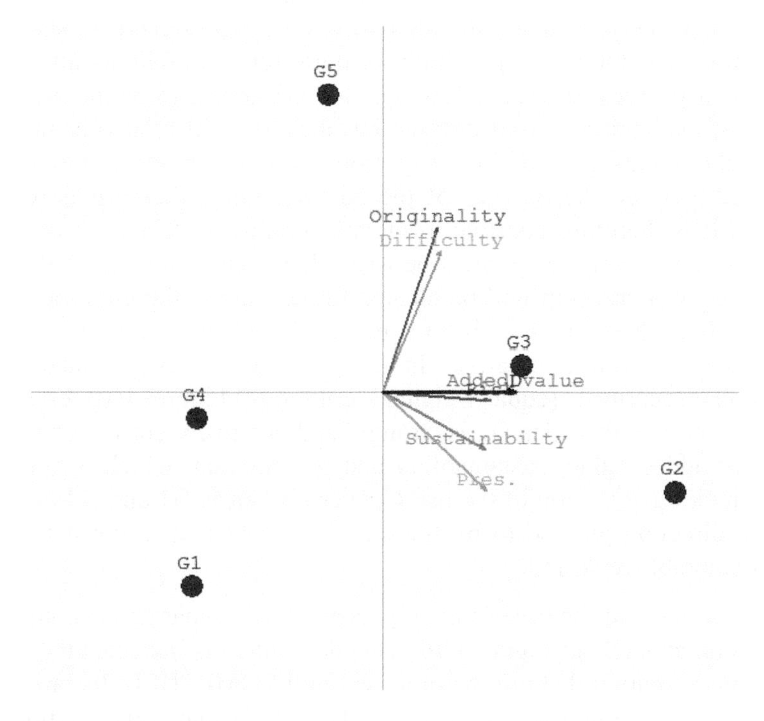

Figure 1.10. *Gaia representation of all ideas and criteria. For a color version of this figure, see www.iste.co.uk/enjolras/decisionmaking.zip*

The criteria are represented by vectors. Both the orientation and the length of these axes are important:

– The axes oriented in the same direction represent criteria that are generally in agreement with each other: for example, the originality criterion and the difficulty of implementing the idea seem to be positively correlated. Conversely, the axes oriented in opposite directions correspond to antagonistic criteria. Thus, one can easily identify criteria that can lead to potential conflicts to be resolved. In our case, and according to Figure 1.10, the postulate of non-correlation of evaluation criteria may need to be reviewed because, for example, the originality and the difficulty of implementing ideas seem to be strongly related. Thus, it would seem that the most original ideas are also those that are most easily implemented.

– The longest axes correspond to the criteria for which the largest deviations between alternatives were observed. A criterion where all alternatives have the same performance, according to the preference function, will have a very short axis in the Gaia plane. As a result, such a criterion will not have a very significant impact within the PROMETHEE rankings. On the other hand, the alternatives are represented by shapes (in our example, by dots). The proximity of these forms highlights alternatives with similar profiles. In Figure 1.10, G2 and G3 appear to be the ideas with the closest evaluations on all criteria, but they are still relatively far apart. Conversely, points placed far away from each other show a noticeable difference between the profiles of the corresponding alternatives (e.g. G1 and G5). It is therefore possible to identify groups or "clusters" of similar actions or, conversely, disparities between them. The alternatives that rank well for a given criterion will be located furthest along the direction of that criterion. It is therefore possible to graphically identify the strengths and weaknesses of each alternative. In our case, if we place ourselves in evaluation scenario 1 (equal-weighted criterion), Figure 1.10 effectively illustrates the fact that idea G2 is strongly and positively correlated with the criteria of added value, sustainability and presentation, which is consistent with its ranking at the top of the list. Conversely, ideas G1 and G4 are in the opposite direction pointed to by the criteria, which is also consistent with their unfavorable ranking.

– The weights of the criteria are represented by a separate axis, known as the decision axis (D) in Figure 1.10. This decision axis indicates the type of compromise solution that will be proposed by PROMETHEE. Its orientation highlights the predominant criteria, and the criteria that can be ignored. In our case, the criteria are equally weighted (scenario 1), so the vector D is positioned at the center of the set of vectors.

Finally, it is important to note that, as in a PCA analysis, although the Gaia plane offers the best possible representation of two-dimensional data, some of the information is lost during this process. The greater the amount of information lost, the less reliable the Gaia plane becomes, and the less useful it becomes for decision makers. To indicate the quality of the Gaia plane, the percentage value is always indicated in the Gaia plane window: this value measures the percentage of information preserved during the calculation of the Gaia plane. In practice, a value greater than 70% indicates a reliable Gaia plane; a value less than 60% means that the interpretation of the Gaia plane must be done with caution.

1.4.2. *Regarding the different versions of PROMETHEE*

The first version of the PROMETHEE I method uses these two flows to rank the alternatives two by two. One alternative outperforms the other if:

– its positive flow is higher and its negative flow is lower;

– its positive flow is higher, and its negative flow is equal;

– its positive flow is equal and its negative flow is lower.

If the positive flow and the negative flow are the same, this means that the alternative neither outranks nor is outranked by the other, and therefore the two alternatives are exactly equal. This problem arises when the positive and negative flows of an alternative are larger than the flows of another alternative because this means that we validate the outranking hypothesis for the positive flow, but at the same time we refute it for the negative flow. Similarly, a similar situation arises if the two streams are more reliable because the upgrade hypothesis is then neither confirmed nor rejected. In these two cases, we have a problem of incomparability. For this, Brans has proposed a related method named PROMETHEE II, which allows users to obtain a complete ranking by avoiding the problems of incomparability. An additional calculation to determine the difference between the positive flow and the negative flow makes it possible to define a net flow associated with each alternative. In this way, by ordering the alternatives according to their net flow, we obtain the final ranking. This is the method we used in the previous section.

Many other versions of the method have been developed: PROMETHEE III for working with intervals, PROMETHEE IV for dealing with a group of continuous solutions, PROMETHEE V for integrating restrictions, PROMETHEE VI providing a representation of the human brain, PROMETHEE GDSS for group decision-making, PROMETHEE Tri which is especially for group sorting problems, and PROMETHEE cluster for classification.

1.5. The PROMETHEE method: instructions for use

1.5.1. *PROMETHEE step by step*

For a simple application of the PROMETHEE method in an Excel spreadsheet, this practical guide offers six steps to follow:

– construction of comparisons between alternatives;

– evaluation of degrees of preference using the preference functions;

– calculation of preference values;

– construction of the aggregate matrix of preferences;

– calculation of incoming and outgoing flows;

– calculation of net flows and final ranking of the alternatives.

To illustrate this step-by-step application, we will apply the method, this time as part of the study of a project to implement a new waste collection system in a suburban city in France. The project has two main objectives: to optimize the logistics network for waste collection, as well as to generate a new source of energy from the treatment of this waste.

The mayor of the city has asked for your help to make a decision that will be best for the city's residents while also respecting the technical and financial constraints of the city.

The company responsible for the implementation of the project has offered five separate alternatives for the new system. An initial classification was carried out based on the available budget and the implementation time of the technical solution (the mayor wanted the solution to be operational before the end of his term).

As a result, three proposals have been selected and will be submitted to the residents. The following criteria have been defined:

– *safety/reliability of the technical solution*: this criterion evaluates the technical aspects of the implementation of the project and the solution in operation, operational safety, technology maturity, maintenance, etc.;

– *energy produced*: each alternative proposes an energy production capacity according to the type of treatment used for the waste; this criterion seeks to determine to what extent this energy can be reused in the city on a daily basis;

– *impact on the current organization of the city*: the implementation of the proposed system will require adjustments to the current infrastructure of the city. These impacts can come in the form of the construction of treatment plants, road closures, etc. This system will also require a cultural change (e.g. if new waste collection points or new ways of recycling are identified). These impacts are seen as having a negative effect, as something to be minimized.

	Reliability	Energy produced (MW)	Impact on the city
Proposal A	8	550	7
Proposal B	3	700	3
Proposal C	5	750	6

Table 1.3. *Evaluation of proposals using the evaluation criteria*

Reliability is evaluated using a scale from 1 to 10 (10 being the most reliable), the energy produced in MW per year, and the impact on a scale from 1 to 10, where 10 represents the highest impact (modification of the current operations of the city, change of habits, visual pollution, etc.).

CAUTIONARY NOTE 1.– The definition of the scales for evaluating the criteria is crucial, as it makes it possible to translate the requirements of the decision maker in a more or less accurate way. A scale from 1 to 5 will be more generic, while a scale from 1 to 10 will make it possible to be more accurate in the evaluations, and to better translate the differences between the alternatives when possible. Having exact values, like the ones related to the energy produced, make it possible to remove all subjectivity from the point at which the value entered is meticulously verified and justified. Finally, it is essential to keep the objective of the criterion in question in mind, as this directly impacts the meaning of the evaluation scale that is used. For example, the impact on the city is rated between 1 and 10, where 10 reflects the highest impact. Thus, the goal is to minimize this criterion, in order to minimize the impact. By contrast, reliability is sought to be maximized.

Finally, a citizens' council was set up to define the priorities of the inhabitants and the town hall regarding the choice of solutions to be put in place. The criteria defined earlier have therefore been weighted in such a way as to reflect their importance. This citizens' council considers that the highest priority is to limit the impact on the city. After that, the reliability of the solution is an important concern. Finally, the energy produced places last. The weightings assigned to the criteria are shown in Table 1.4.

Criterion	Reliability	Energy produced	Impact on the city
Weighting	25%	15%	60%

Table 1.4. *Weighting of criteria*

Based on all of these data, you choose to use PROMETHEE to carry out the ranking of the alternatives. During a working session with the citizen's council, you were able to build the preference functions for the three criteria taken into account.

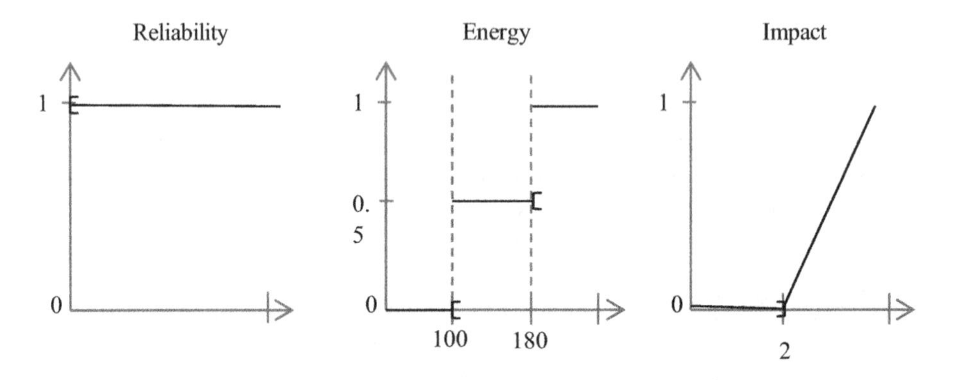

Figure 1.11. *Preference functions for reliability, energy, and impact criteria*

CAUTIONARY NOTE 2.– Preference functions should be created with decision makers in order to best translate their preferences. They can possibly be build relying on verbal declarations. For example, if a decision maker says: "I always prefer the most reliable alternative, regardless of the difference between the two systems", this translates into a preference function for the criterion of "reliability", as above. On the other hand, if a decision maker says: "for a difference in energy production of less than 100 MW, I think that the systems are equivalent in terms of performance", this is a threshold of indifference q of 100 MW for the energy criterion produced, as shown in Figure 1.11.

1.5.1.1. *Construction of comparisons between alternatives*

For each pair of alternatives, construct a comparison matrix as follows:

– compare the two values for each criterion;

– enter the value of the difference in the column next to the preferred solution.

For example, for the reliability criterion, alternative A is preferred, and the value is higher. I write "5" on the left part of the table.

CAUTIONARY NOTE 3.– Pay attention to the criteria that must be minimized. In many cases, the impact must be minimized, and therefore the lowest impact values are preferred.

Alternative A		Criterion	Alternative B	
Difference	Value		Value	Difference
5	8	Reliability	3	
	550	Energy	700	150
	7	Impact	3	4

Table 1.5. *Comparison matrix between alternatives*

Repeat the same operation and build the same comparison matrices between alternatives AC and BC.

1.5.1.2. *Evaluation of levels of preference P(i,j) using the preference functions for each criterion*

For the reliability criterion, its related preference function means that for any difference between two alternatives, the preference for the most reliable alternative is a total preference (and therefore equal to 1).

For the energy produced criterion, for any difference less than 100, the preference is zero (we are below the threshold of indifference), for a difference between 100 and 180, the difference is intermediate (0.5) for the alternative producing the most energy, and finally, for a difference between two alternatives greater than 180, the alternative producing the most energy is totally preferred (preferably equal to 1).

For the impact criterion, the indifference threshold is placed at 2. Below this deviation value, the preference is zero. On the other hand, after this value, the preference is linear, and therefore calculated according to the following formula:

$$P_i(X,Y) = \frac{\text{difference}_i(X,Y)}{\text{difference}Max_i},$$

with:

– X and Y being the two alternatives considered;

– i being the criterion considered;

– $\text{difference}_i(\text{X}, \text{Y})$ being the value of the difference between the alternatives X and Y for criterion i;

– $\text{difference}Max_i$ being the maximum difference existing between all the alternatives for criterion i.

Therefore, $P_{impact}(B,A) = \frac{\text{difference}(B,A)}{differenceMax_{impact}} = \frac{4}{4} = 1.$

Alternative A			Criterion	Alternative B		
$P(A,B)$	Difference	Value		Value	Difference	$P(\text{B},\text{A})$
1	5	8	Reliability	3		0
0		550	Energy	700	150	0.5
0		7	Impact	3	4	1

Table 1.6. *Differences and preferences between alternatives A and B*

Alternative A			Criterion	Alternative C		
$P(A,C)$	Difference	Value		Value	Difference	$P(C,A)$
1	3	8	Reliability	5		0
0		550	Energy	750	200	1
0		7	Impact	6	1	0

Table 1.7. *Differences and preferences between alternatives A and C*

Alternative B			Criterion	Alternative C		
$P(B,C)$	Difference	Value		Value	Difference	$P(C,B)$
1		3	Reliability	5	2	1
0		700	Energy	750	50	0
0		3	Impact	6	3	0.5

Table 1.8. *Differences and preferences between alternatives B and C*

1.5.1.3. *Calculation of preference $\pi(a_i, a_j)$values for each comparison (see equation [1.2])*

$$\pi(a_i, a_j) = \sum_{k=1}^{n} w_n P_k \left(f_k(a_i) - f_k(a_j)\right),$$

with:

– n, number of criteria;

– ω_n, the weighting of each criterion n;

– $P_k \left(f_k(a_i) - f_k(a_j)\right)$ the degree of preference from a_i to a_j.

In our case:

$$\pi(A,B) = ((0.25 \times 1) + (0.15 \times 0) + (0.6 \times 0)) = 0.25,$$

$$\pi(B,A) = ((0.25 \times 0) + (0.15 \times 0.5) + (0.6 \times 1)) = 0.675.$$

CAUTIONARY NOTE 4.– In PROMETHEE, the preference values are not symmetrical due to the impact of preference functions (e.g. above, the AB and BA comparisons give different results). You must therefore repeat the same calculation for all combinations.

1.5.1.4. *Aggregation of all values of π in a matrix*

$\pi(a_i, a_j)$	A	B	C
A	0.00	0.25	0.25
B	0.68	0.00	0.30
C	0.15	0.25	0.00

Table 1.9. *Preference values between alternatives*

1.5.1.5. *Calculation of incoming and outgoing flows*

Calculate the flows $\phi^+(a_i)$ and $\phi^-(a_i)$, as the sums of rows and columns divided by the number of alternatives minus one (see equations [1.3] and [1.4]), as well as the net flow $\phi(a_i)$ for each alternative (equation [1.5]).

$\pi(a_i, a_j)$	A	B	C	$\phi^+(a_i)$
A	0.00	0.25	0.25	0.25
B	0.68	0.00	0.30	0.49
C	0.15	0.25	0.00	0.20
$\phi^-(a_i)$	0.41	0.25	0.28	
$\phi(a_i)$	–0.1625	0.2375	–0.075	

Table 1.10. *Calculation of flows $\phi^+(a_i)$, $\phi^+(a_i)$ and net flow $\phi(a_i)$*

1.5.1.6. *Final ranking of alternatives*

As a final step, suggest the ranking of alternatives. The higher the value of $\phi(a_i)$, the more favorably the alternative is classified. In our case, the ranking is B, C and A.

1.5.2. *Application of PROMETHEE with software support*

The previous section presents an application of PROMETHEE done "manually", allowing for results to be obtained that are interesting but rather static. Now, as was noted in the example related to the selection of ideas, the results are often sensitive to changes that are related to the evaluation strategies chosen (mainly the weightings associated with the criteria). Thus, a dynamic and interactive view of these results can strengthen the analysis that is performed. In addition, other graphical tools such as the Gaia plane can be valuable assets to interpret the rankings that are obtained and progress toward making the most informed decision possible.

Thus, we will focus in this section on the Smartpicker® software, which is a multi-criteria analysis tool based on PROMETHEE for classifying alternatives based on degrees of preference (Ishizaka and Nemery 2013). The main advantage of this software is that it allows for dynamic analyses to be done on the results obtained, presented in a visual and interactive way.

Let us look at each of the steps the case dealt with in the previous section on waste management in a city in France. The first task is to load the initial data into the software, either by importing it through an Excel table or manually. The interface is illustrated in Figure 1.12 and represents the matrix for evaluating the alternatives according to the chosen criteria.

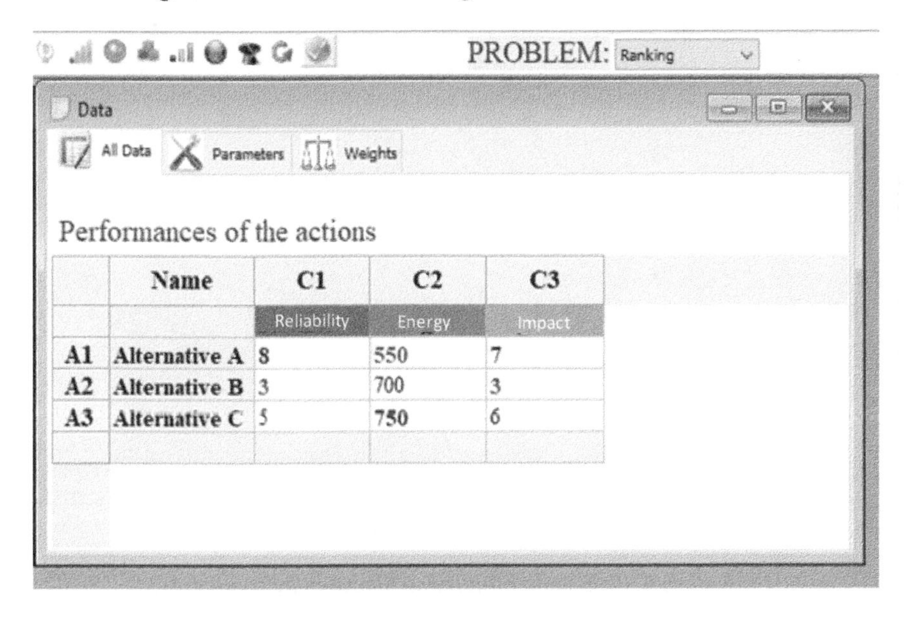

Figure 1.12. *Entry of criteria and alternatives. For a color version of this figure, see www.iste.co.uk/enjolras/decisionmaking.zip*

The next step is to define the parameters, that is, to build the preference functions for each criterion (Figure 1.13). To accomplish this, a dynamic interface will allow you to choose the type of preference function and to define the thresholds of indifference and preference. In the example above, we illustrate the construction of the preference function related to the criterion "impact on the city". We also note that this interface makes it possible to choose the objective associated with the criterion (here, we are trying to minimize this criterion). This information is crucial for the related flow calculation.

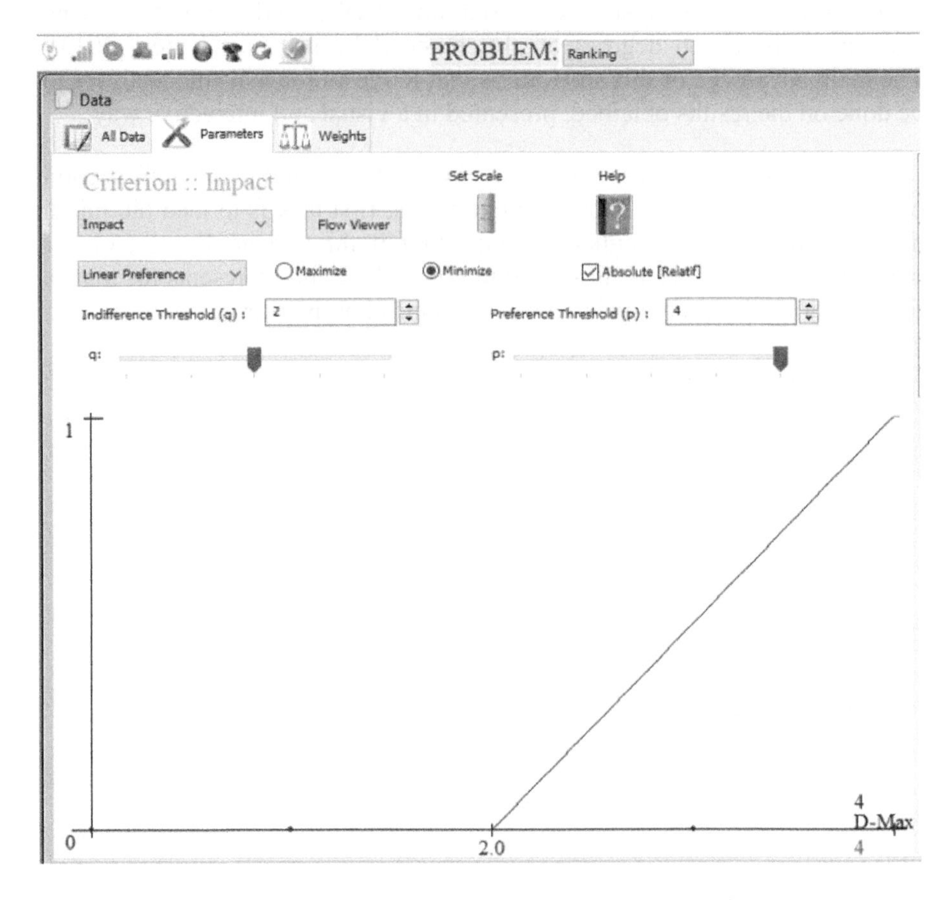

Figure 1.13. *Setting preference functions by criteria*

Finally, the last data to consider are the weighting criteria (Figure 1.14). This must also be integrated through a dedicated tab. Thus, the weightings obtained during the citizen's council are added to the analysis.

The next step is to view the ranking alternatives put forward by the software. The results are presented in the form of histograms and allow for different complementary analyses. Among other things, the "descriptive criteria" feature allows users to highlight the contribution of each criterion in the calculation of the net associated with each alternative (Figure 1.15). This makes it possible to visualize the strengths and weaknesses of each alternative and explain the reasons for where it was placed in the ranking. In particular, this functionality was used in section 1.3.1.4 to analyze the results of our application case.

It is important to clarify that the ranking that is obtained remains dynamic, in the sense that it is possible to modify parameters such as the weight of the criteria, and to instantly visualize the impact of these changes on the ranking. This makes it possible to test the sensitivity of the results and explore different scenarios.

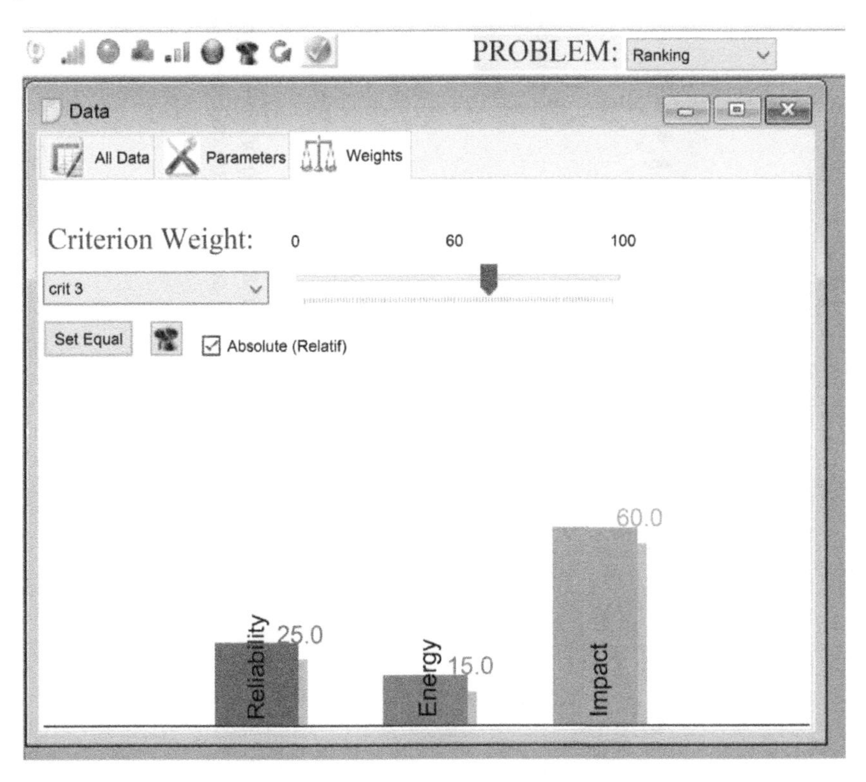

Figure 1.14. *Assignment of weightings of criteria*

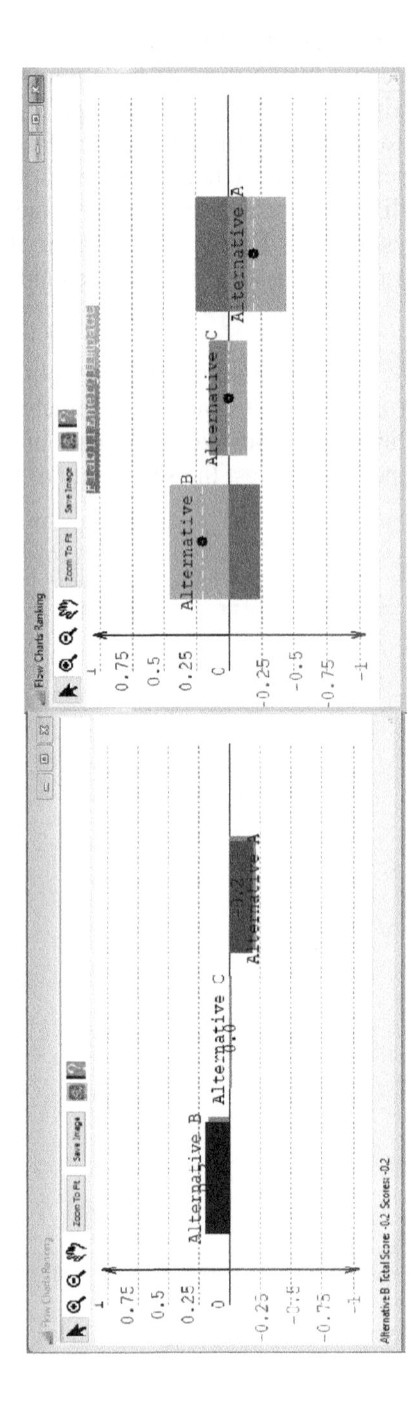

Figure 1.15. *Visualization of the ranking of alternatives and description. For a color version of this figure, see www.iste.co.uk/enjolras/decisionmaking.zip*

Finally, a Gaia plane can be generated to complete the analysis of the results (Figure 1.16). It will represent the “criteria” vectors and the “alternative” points, and make it possible to visualize the correlations between criteria, or the similarities/disparities between alternatives. Vector D, in black, reflects the preferences of the decision maker. It is directly dependent on the weights assigned to the criteria. The reliability of the Gaia plane is also indicated at the bottom of the graph. In this example, 100% of the data are represented in the chosen plane, which means it is very reliable in terms of interpretation.

This Gaia plane is also interactive at different levels (Figure 1.17). First of all, a feature makes it possible to carry out a projection of the alternatives on the decision vector D. This feature makes it possible to approximately reproduce the ranking obtained through the calculation of the net flow. The best alternative is thus the one whose orthogonal projection is closest to the tip of the vector D. In this example, we obtain the classification B > C > A, which is consistent with the results obtained previously.

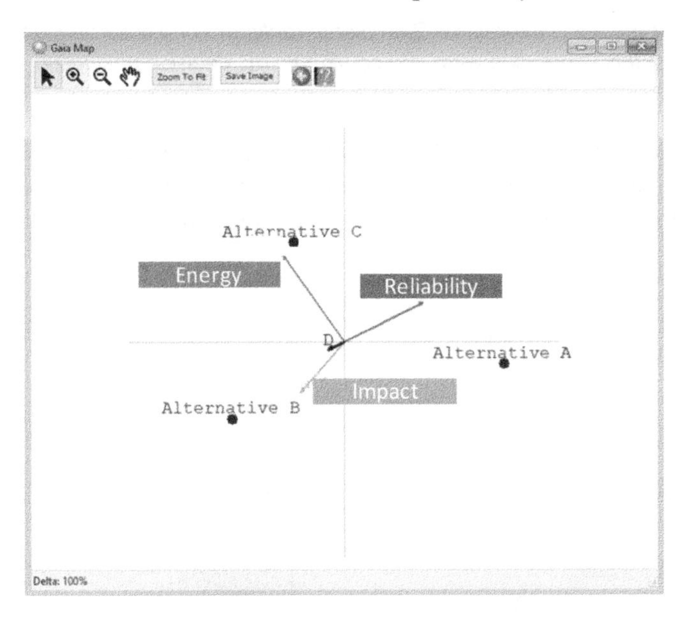

Figure 1.16. *Gaia plane. For a color version of this figure, see www.iste.co.uk/enjolras/decisionmaking.zip*

On the other hand, as is the case for the ranking of the alternatives using histograms, it is possible to modify the value of the weights assigned to the criteria and to instantly visualize how the decision vector D (as well as the ranking) changes.

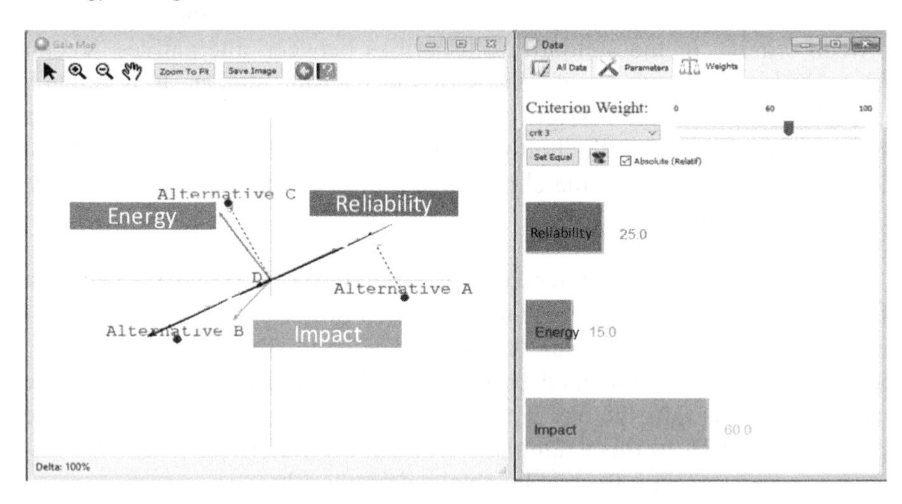

Figure 1.17. *Interactive Gaia plane: ranking. For a color version of this figure, see www.iste.co.uk/enjolras/decisionmaking.zip*

In the example below, we have assigned the same weight value to each criterion. The change in the orientation of the vector D, as well as the classification, is clearly visible. This gives Alternative C as coming out on top (Figure 1.18).

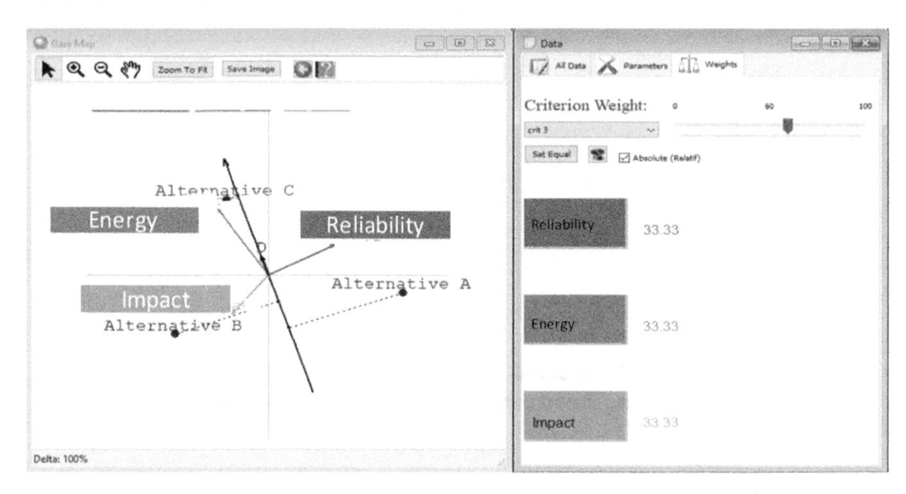

Figure 1.18. *Interactive Gaia plane: change on the basis of the weightings*

As we can see, this software offers an advanced application of the PROMETHEE method. In particular, it makes it possible to carry out complementary and dynamic analyses and provides a more accessible interface for decision makers. This makes the approach more interactive and encourages the exploration of alternative scenarios.

PROMETHEE AT A GLANCE.– The Objective is that the PROMETHEE method carries out a comparison between alternatives through paired groupings, using the preference function in order to determine the best one.

Its defining characteristic is that in addition to its ease of use, it makes it possible, when necessary, to avoid the phenomenon of compensation between criteria (which is unique among the total aggregation methods: MAUT, AHP and weighted average).

Its limitations are that it does not take into account the interactions between criteria. On the other hand, the degree of importance (weighting) of the criteria must be known beforehand.

1.6. References

Acar, S. and Runco, M.A. (2012). Creative abilities: Divergent thinking. In *Handbook of Organizational Creativity*, Mumford, M.D. (ed.). Academic Press, San Diego.

Brans, J.P. and Mareschal, B. (1994). The PROMCALC & Gaia decision support system for multi-criteria decision aid. *Decis. Support Syst.*, 12, 297–310.

Brans, J.P. and Vincke, P. (1985). A preference ranking organisation method. *Manag. Sci.*, 31, 647–656.

Carson, S.H., Peterson, J.B., Higgins, D.M. (2003). Decreased latent inhibition is associated with increased creative achievement in high-functioning individuals. *J. Pers. Soc. Psychol.*, 85, 499–504.

Correa, C.H. and Danilevicz, Â.D.M.F. (2015). Method for decision making in the management of innovation: Criteria for the evaluations of ideas. In *International Association for Management of Technology*, Cape Town, South Africa, 8–11 June, 2151–2169.

Craven, C.E. (2015). Refusing to be toured: Work, tourism, and the productivity of "life" in the Colombian Amazon. *Antipode*, 48, 544–562.

De Bono, E. (2010). *Lateral Thinking: A Textbook of Creativity*. Penguin, London.

De Brabandere, L. (2002). *Le management des idées : de la créativité à l'innovation*. Dunod, Malakoff.

Eberle, R.F. (1972). Developing imagination through scamper. *J. Creat. Behav.*, 6, 199–203.

Gabriel, A., Camargo, M., Monticolo, D., Boly, V., Bourgault, M. (2016a). Improving the idea selection process in creative workshops through contextualisation. *J. Clean. Prod.*, 135, 1503–1513.

Gabriel, A., Monticolo, D., Camargo, M., Bourgault, M. (2016b). Creativity support systems: A systematic mapping study. *Think. Ski. Creat.*, 21, 109–122.

Gabriel, A., Monticolo, D., Camargo, M., Bourgault, M. (2017). Conceptual framework of an intelligent system to support creative workshops. In *TRIZ – The Theory of Inventive Problem Solving*, Cavallucci, D. (ed.). Springer, Berlin.

Howard, T.J., Culley, S.J., Dekoninck, E. (2008). Describing the creative design process by the integration of engineering design and cognitive psychology literature. *Des. Stud.*, 29, 160–180.

Kudrowitz, B.M. and Wallace, D. (2013). Assessing the quality of ideas from prolific, early-stage product ideation. *J. Eng. Des.*, 24, 120–139.

Nemery, P., Ishizaka, A., Camargo, M., Morel, L. (2012). Enriching descriptive information in ranking and sorting problems with visualizations techniques. *J. Model. Manag.*, 7, 130–147.

Obando Lugo, J., Ochoa, F., Fredy, A., De Duque, R., Isabel, R., Rozo, E., Villada, I. (2010). Enfoque metodológico para la formulación de un sistema de gestión para la sostenibilidad en destinos turísticos. *Anu. Tur. Soc.*, 11, 175–200.

Ochoa, G. (2008). El turismo ¿una nueva bonanza en la Amazonia? In *Fronteras en la globalización: localidad, biodiversidad y comercio en la Amazonia*, Zárate, C. and Ahumada Beltrán, C. (eds). Observatorio Andino, Pontificia Universidad Javeriana, Bogotá.

Osborn, A.F. (1963). *Applied Imagination; Principles and Procedures of Creative Problem-Solving: Principles and Procedures of Creative Problem-Solving*. Scribner, New York.

Riedl, C., Blohm, I., Leimeister, J.M., Krcmar, H. (2010). Rating scales for collective intelligence in innovation communities: Why quick and easy decision making does not get it right. In *Proceedings of the International Conference on Information Systems*. ICIS, Saint-Louis, 12–15.

Sawyer, K. (2012). *Explaining Creativity: The Science of the Human Innovation*. Oxford University Press, New York.

Stevens, G.A. and Burley, J. (1997). 3,000 raw ideas = 1 commercial success! *Res. Technol. Manag.*, 40, 16–27.

Van Gundy, A.B. (1987). *Creative Problem Solving: A Guide for Trainers and Management.* Quorum Books, Westport.

Verhaegen, P.-A., Vandevenne, D., Peeters, J., Duflou, J.R. (2013). Refinements to the variety metric for idea evaluation. *Des. Stud.*, 34, 243–263.

Westerski, A. (2013). Semantic technologies in idea management systems: A model for interoperability, linking and filtering. Thesis, Universidad Politécnica de Madrid, Escuela Técnica Superior de Ingenieros de Telecomunicacion, Madrid.

2

The Upstream Phases of Product Design: An Application of AHP

The development of an innovation always begins with an idea that must then be successfully implemented. Design is the term used for all activities that bring ideas to life, progressing through different stages of materialization. Concepts, models, specifications, prototypes and user tests are all intermediate projects that lead to a viable final product. This makes design an essential component of innovation capacity. It requires human abilities, the right type of organization and technical and methodological means, and it can be facilitated by the use of digital tools. However, the process of designing an innovation is filled with uncertainties that must be sorted out progressively. Thus, decision-making in this context involves risks, and can have strong impacts on the costs, duration and acceptability of the final product, especially during the early design phases.

To illustrate this problem, we will examine a case study proposed by Serna et al. (2016) in the chemical industry. Specifically, we will be looking at the notion of sustainability in the formulation of a chemical product: ethylacetate. This article proposes an application of the analytical hierarchy process (AHP) method developed by Thomas Saaty in the 1970s.

2.1. Context and challenges in decision-making

2.1.1. *The challenges of the upstream phases of the design process*

Design plays a role that brings together the activity of creativity and the implementation of innovations on the market. It seeks to develop an innovative product that is easily and quickly marketable, minimizing the design failure rate (Fabbri 2017). For designers as well as clients, it is necessary to materialize, illustrate and test ideas to give them legitimacy and adapt them to the constraints and needs of the organization and the market. This allows them to be shaped by considering the needs and requirements of customers, and then to make them testable and communicable through prototypes. This step makes it possible to assess the potential of the idea, identify the main defects that need to be solved and begin to determine the manufacturing costs. Thus, having accessible, easy-to-use and varied technical resources to carry out this prototyping phase is an essential factor to facilitate exchanges with other people, stimulate creative rebounds and promote the development of projects. It is also essential to bring together mixed project teams, made up of members from different departments within the company (R&D, marketing, design, purchasing, legal, etc.). This multidisciplinary approach allows the organization to acquire new knowledge and skills for the purpose of removing technical, administrative or legislative barriers. This approach can also lead to opportunities for new projects. The involvement of stakeholders from outside the organization (customers, suppliers, makers, etc.), as well as the use of collaborative design and manufacturing workshops/events (hackathons, user testing, etc.), also seek to accelerate the innovation process. In the era of digitalization, digital tools (simulations, virtual reality, collaborative tools, etc.) as well as new materialization methods (3D printing, rapid prototyping, etc.) play a crucial role in identifying development opportunities, accelerating product development, limiting the cost of real trials and facilitating collaborative work. In this way, they are changing the way companies manage their design process.

In fact, in recent years, we have seen a steep decline in the lifespan of products, which become obsolete faster year after year. At the same time, the time required for their adoption by users has also been drastically reduced.

Thus, while it took decades for products/technologies such as the automobile, the telephone, or the credit card to reach 50 million users, in recent times it has only taken Facebook and Twitter 2 or 3 years to reach this same level. The main interest in this reduction is both economic and strategic, because by combining the attractiveness related to the novelty of the product and the high financial margin that can result from it, a strong competitive advantage is created. However, this reduction in the lifecycle implies a multiplication of projects, each of which has its own needs in terms of development times.

In this context, the early design phases become crucial, as they are the ones that include the most uncertainties, but that also have the most impact on the process as a whole. The weight of the first decisions is thus very important. Indeed, during these upstream phases, the process is not yet well defined. This means every decision is crucial because it can strongly impact the entire process. It is thus necessary to simultaneously take into account different options and dimensions that sometimes conflict, so as to limit costs, respect deadlines and improve the performance of the product as such. The value of multi-criteria analyses is thus obvious at this stage of the innovation process: to support complex decisions under an uncertain environment, which can have a strong impact later on.

2.1.2. *A trend in innovation: sustainable design*

In recent years, we have seen the explosion of a number of strong trends, such as the rise of the collaborative economy, the energy transition or the development of new digital technologies, which influence the design phase of innovation. Whether the nature of these changes is technological or use-related, these trends are proving to be particularly destabilizing for companies (Acosta et al. 2014) and thus require new models of design and organization.

We have chosen to focus primarily on sustainable design. Indeed, for several years now, the notion of sustainability has been gaining importance. In particular, this has been reflected in new regulations, new industrial practices and advances in scientific research and education. Sustainability is based on three pillars: the economy, the environment and the social realm (Hacking and Guthrie 2008). Though the majority of companies focus on the environmental dimension, which is "easier" to implement (Acosta et al. 2014), sustainable innovation nevertheless extends beyond its ecological dimension, including economic and social approaches such as professional reintegration, participatory innovation, circular economy patterns, recycling.

Companies that are committed to a sustainable design approach thus have to make trade-offs, because operationally speaking, the importance of these three pillars in decision-making varies according to the design processes that are considered and according to the context in which they are carried out. However, it is difficult to determine this importance, especially in the early phases of the design process, which are filled with uncertainties. Therefore, we will illustrate this problem of sustainable design through a case study in the chemical industry. By proposing a weighting method of the pillars of sustainable development, this example reflects their importance in the choice of the formulation of a chemical product.

2.1.3. *Definition of the decision model*

The case we are going to study is inspired by the article by Serna et al. (2016), proposing a methodology for selecting sustainable chemical processes during the early stages of chemical formulation. The chemical process is defined as the collection of raw materials and the sequence of chemical reactions that makes it possible to convert them into the desired product. Different chemical process options can be used to create the same product. Therefore, the choice between one or the other of these chemical processes will set the guideline of the design process. This is a key decision in the particular context of the chemical industry.

Serna et al. (2016) apply their methodology to the case of the production of ethylacetate. Specifically, the objective of this article is to select the most durable chemical process to produce this molecule, which is mainly used as a solvent for the manufacture of resins, inks, herbicides or food products, among other things.

Four different chemical processes are considered, all of which are used to create the desired molecule: ethylacetate.

$$CH_3-C(=O)-OH + CH_3CH_2OH \rightleftharpoons CH_3-C(=O)-OCH_2CH_3 + H_2O$$

Reaction 2.1. *Fischer esterification*

$$2\ CH_3CH_2OH \longrightarrow CH_3-C(=O)-OCH_2CH_3 + 2\ H_2$$

Reaction 2.2. *Dehydrogenation of ethanol*

$$2\ CH_3-C(=O)-H \longrightarrow CH_3-C(=O)-OCH_2CH_3$$

Reaction 2.3. *Condensation of acetaldehyde*

$$H_2C{=}CH_2 + CH_3-C(=O)-OH \longrightarrow CH_3-C(=O)-OCH_2CH_3$$

Reaction 2.4. *Direct addition*

In order to assess the sustainability of each of these processes (reactions 1–4), Serna et al. (2016) relied on the three pillars of sustainable development in order to define a benchmark of relevant criteria: the economy, the environment and social aspects.

Four criteria have been defined, each presenting several sub-criteria making it possible to characterize them. These sub-criteria are evaluated on the basis of the properties and operating conditions of the chemical reactions considered.

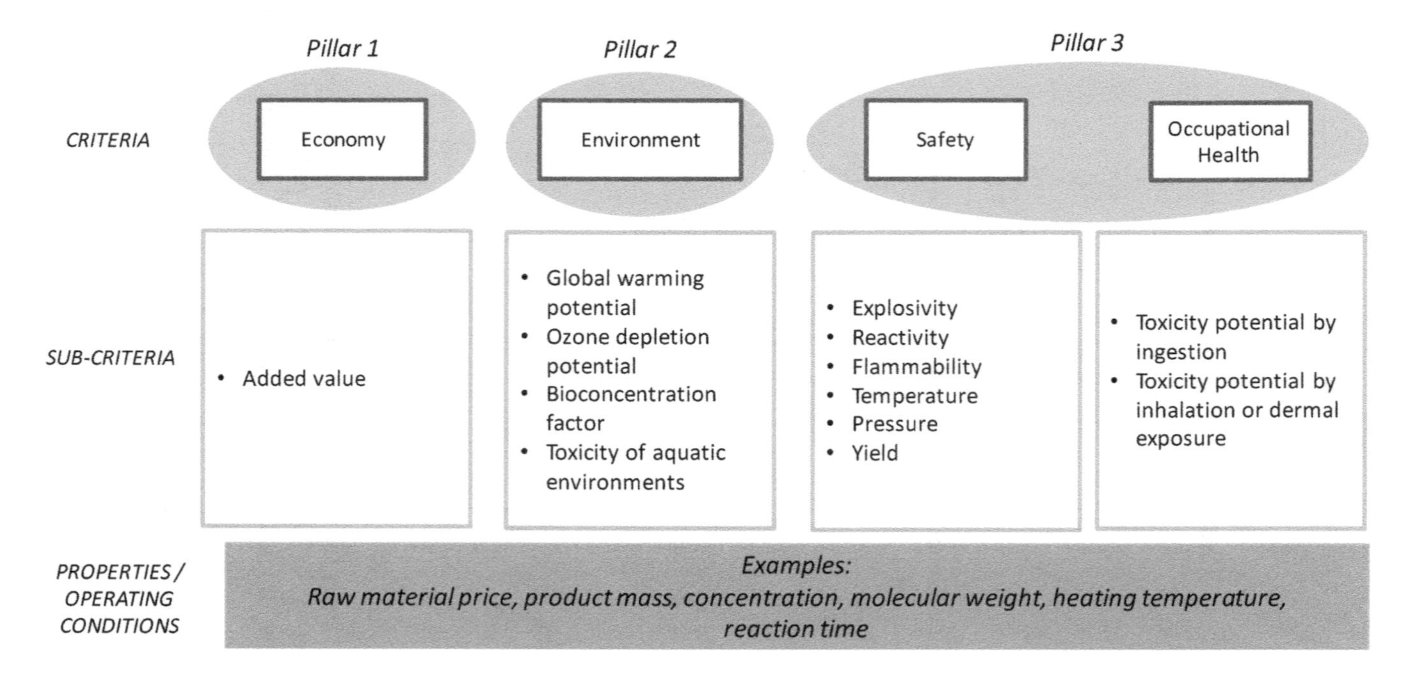

Figure 2.1. *Decision model*

Thus, the proposed decision model is a hierarchical model with three levels: criteria, sub-criteria and chemical properties/operating conditions, allowing for the selection of sustainable chemical processes during the early phases of the formulation of a chemical product.

2.2. The AHP method: analytic hierarchy process

The AHP method was developed by Thomas Saaty in the 1970s. It is the best known and most commonly used multi-criteria analysis method. It is a ranking-based method, which allows alternatives to be ranked based on different evaluation criteria ranging from the "best" to the "worst" from the point of view of decision makers. The biggest defining feature of AHP is that it breaks the problem into a hierarchical structure at several levels (criteria/sub-criteria/indicators, etc.) and allows criteria to be weighted according to their importance in decision-making (Saaty 1980). This weighting process is based on two fundamental principles: the pairwise comparison of criteria and the validation of the consistency of these comparisons.

2.2.1. *The fundamental principle: the relative importance of criteria*

The AHP method assumes that not all the criteria considered in a decision problem necessarily have the same importance. Some of them will have a stronger impact and will be weighted differently in the balance used for the final decision. This is directly related to the point of view of the decision makers, who will intuitively pay more attention to some criteria than others. Taking into account their relative importance, characterized by *the concept of weighting*, the preferences of decision makers are modeled with somewhat greater precision by refining the framework of the evaluation criterion.

The specific strength of AHP is that, using this method, we do not consider the weighing criteria as raw input data, but instead as a fully separate construction stage of the decision problem. The creation of these weightings can in fact be carried out directly by decision makers. For example, you start with 100 points to distribute over these four evaluation criteria. Which ones are the most important to you? However, this direct method can be made more accurate and objective by using an appropriate approach. To this end, AHP includes a stage for the weighting of criteria based on *the principle of pairwise comparisons*.

This step makes it possible to obtain weighted criteria based on the opinions of decision makers using the Saaty scale (Ghazinoory et al. 2007). The Saaty scale allows for the importance of one criterion relative to another to be measured by assigning a numerical value to each comparison. The scale is progressive: a value of 1 means that the two criteria are of "equal importance", and the value 9 means that one criterion is "very much more important" than the other (Saaty 2008). Thus, the values 1, 3, 5, 7 and 9 are associated with a corresponding linguistic formulation (Table 2.1), allowing decision makers to determine their own position on this scale. It is also possible to use the intermediate values of 2, 4, 6 and 8 to gain accuracy in comparisons when necessary. The objective is therefore to carry out all the combinations of criteria in such a way as to compare them with each other.

Saaty scale	Corresponding linguistic formulations
1	Of equal importance
3	Slightly more important
5	Moderately more important
7	Much more important
9	Very much more important

Table 2.1. *Saaty scale (Saaty 1980)*

Here, we will use the classic example of buying a car as a decision-making case. The criteria considered for making a decision on which car to buy are price, power and comfort. The resulting pairwise comparisons for this are shown in Figure 2.2.

In this example, the set of possible combinations can be divided into three pairs of criteria. By making these comparisons one by one, the decision maker considers that the price is relatively more important than the power (given a value of 5 on the left side of the Saaty scale). On the other hand, the decision maker finds that price and comfort have the same importance (value 1 on the Saaty scale) and that comfort is slightly more important than power (value 3 on the right side of the Saaty scale). By then combining all of these comparisons, we can then construct a square matrix whose diagonal is equal to 1. This owes to the fact that a criterion when compared to itself will necessarily have the same level of importance. In addition, the values are reversed on each side of the diagonal, in order to represent the two directions of comparison: if the price/power comparison has a value of 5, then the power/price comparison is represented by the multiplicative inverse of the previous value: 1/5. This characteristic of the comparison matrix makes it possible to carry out the analysis of the consistency of the comparisons made.

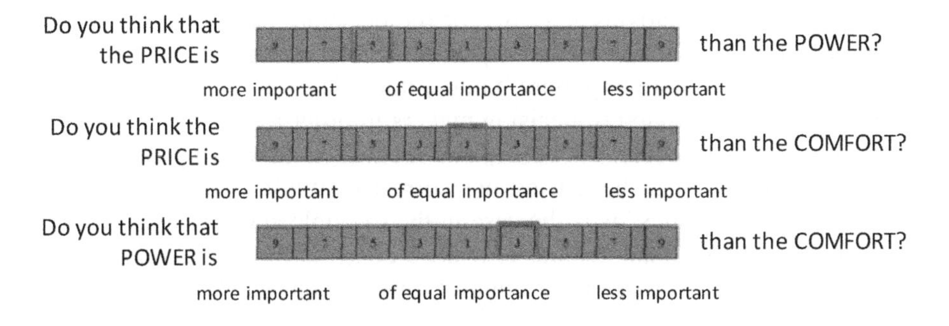

Figure 2.2. *Example of a pairwise comparison of criteria (Saaty scale). For a color version of this figure, see www.iste.co.uk/enjolras/decisionmaking.zip*

	Price	Power	Comfort
Price	1	5	1
Power	1/5	1	1/3
Comfort	1	3	1

Table 2.2. *Pairwise comparison matrix (A)*

Finally, a normalization of this matrix (see section 2.4, "AHP: instructions for use") as well as the determination of the average of each row make it possible to obtain a weight for each criterion, in the form of a percentage (the sum of the weights of the set of criteria is therefore equal to 1).

	Price	Power	Comfort	Weighting
Price	0.45	0.56	0.43	48%
Power	0.09	0.11	0.14	11%
Comfort	0.45	0.33	0.43	41%

Table 2.3. *Calculation of weighting criterion*

The main advantage of these pairwise comparisons is that they make it possible to sub-divide the entire framework of criteria, making it easier to characterize. In fact, it is intellectually easier to compare two elements with each other, rather than four or five simultaneously. Thus, this comparison step appears to be relatively intuitive for the decision maker. In addition, it makes it possible to transform a qualitative assessment ("slightly more important", "much more important") into a numerical value, toward the goal of quantifying the opinion of decision makers in order to be able to model it as effectively as possible. However, this method is limited by the mental load necessary for it to be carried out correctly. While it may be easy to make three comparisons, as is the case in the example above (Figure 2.2), it is much more difficult and tedious to make a dozen or more, especially since it is important to ensure that these comparisons are consistent.

Let us go back to the example given above. In view of the results, it seems that price and comfort have the same importance and that they are both more important than power. It can be interpreted in the following way:

Statement 1: "(price = comfort) > power".

However, if we quantify this according to Table 2.2, we obtain:

Statement 2: "Price = 5*Power".

And Statement 3 "Comfort = 3*Power".

However, according to statement 1, we also obtain:

Statement [4]: "Price = comfort".

This shows that there is an inconsistency in the comparisons, which are carried out with a reduced number of criteria, since statements 2, 3 and 4 are not compatible. While this inconsistency can easily be detected and removed in this basic example, this is not always the case when making larger scale comparisons.

Thus, the AHP method includes a stage for validating consistency comparisons, by calculating a consistency ratio RC. This value is calculated on the basis of a linear algebra problem, aimed at identifying the eigenvectors of the matrix characterizing the pairwise comparisons that have been made (Table 2.2), traditionally called matrix A.

The first step is to calculate λmax, which represents the average of the eigenvalues of the pairwise comparison matrices (A):

$$\text{Eigenvector } (A) = \frac{\vec{A} \cdot \vec{\omega}}{\vec{\omega}},$$

with $\vec{\omega}$ the vector of the weights of the criteria. The value retained for λmax is equal to the average of the values of the eigenvector of (A).

Next, the consistency index IC must be calculated:

$$\text{IC} = \frac{\lambda\text{max} - K}{K - 1},$$

where K is the number of criteria. Finally, the consistency ratio RC is obtained by dividing IC by the random index of Saaty IA. This random index is characterized by standard values, dependent on the number of criteria considered (K). These values can be found in the dedicated table (Figure 2.3) (Saaty 2008).

$$\text{RC} = \frac{\text{IC}}{\text{IA}}.$$

K (Nb of criteria)	2	3	4	5	6	7	8	9	10	11
IA	0	0.58	0.90	1.12	1.24	1.32	1.41	1.45	1.49	1.51

Figure 2.3. *Table of values of the random index IA*

This RC ratio represents the likelihood that the comparisons were made randomly. A high ratio indicates a high level of inconsistency, while a value of 0 represents perfect consistency. However, it is very rare for decision makers to be perfectly consistent in pairwise comparisons, and the greater the number of criteria to be compared, the higher the risk of inconsistency. That is why the number of criteria to be compared is traditionally limited to seven, representing a mental load that can be considered acceptable for the human brain. In addition, the threshold of acceptability of the consistency ratio is set at around 10% (although this may vary slightly, depending on the number of criteria considered) in order to allow a margin of flexibility during pairwise comparisons.

For our example, applying the steps presented earlier, the vector of the eigenvalues of the matrix A obtained is equal to:

$$\text{Eigenvector}\ (A) = \frac{\vec{A}.\vec{\omega}}{\vec{\omega}} = \begin{cases} 3.04 \\ 3.01, \\ 3.03 \end{cases}$$

with $\vec{\omega} = \begin{cases} 0.48 \\ 0.11 \\ 0.41 \end{cases}$ given according to Table 2.3. Thus:

$$\lambda\text{max} = \text{Average [Eigenvector}\ (A)] = 3.03 \text{ and } \text{IC} = \frac{\lambda\text{max}-K}{K-1} = \frac{3.03-3}{3-1} = 0.015.$$

According to the Saaty table, the value of IA for a three-criteria decision problem is $\text{IA} = 0.58$.

And it follows that: $\text{RC} = \frac{\text{IC}}{\text{IA}} = \frac{0.015}{0.58} = 0.03 = 3\%$.

The consistency ratio is therefore acceptable.

Thus, in order to obtain the most accurate model of the preferences of decision makers, the AHP methodology integrates the notion of weighted criteria by proposing a weighting step of such criteria, as well as a validation of the results obtained. This weighting step is crucial for carrying out the method and conditions of the entire decision model.

2.2.2. *Application of the method*

In order to concretely visualize an application of AHP, we will rely on an application case from the study by Serna et al. (2016). The authors propose a methodology that allows for sustainable processes for the formulation of a chemical product to be identified. This methodology is based on a decision model including four criteria, which must be weighted according to their importance in the assessment of sustainability of an alternative that allows for the formulation of ethylacetate. The stages of this methodology are presented in Figure 2.4. These steps, and in particular the third and fourth steps, are based on the AHP multi-criteria method, which we will detail using this case study.

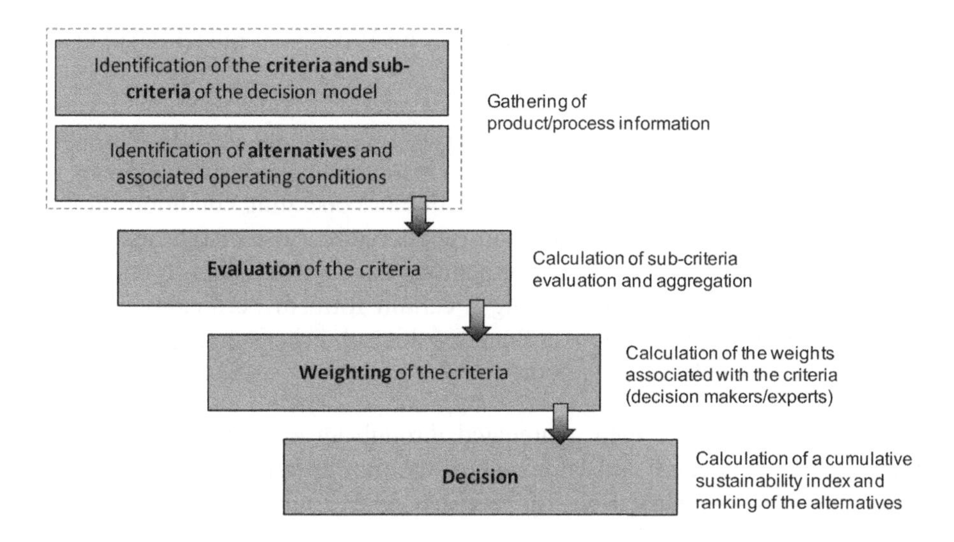

Figure 2.4. *Methodology adopted*

2.2.2.1. *Identification of criteria and alternatives*

This initial step first seeks to build the framework of decision criteria to be considered. In our case, this work was carried out in the construction phase of the decision model, presented by Figure 2.1 (section 2.1.3): four criteria, as well as the corresponding sub-criteria, were identified and described.

The second objective of this step is to characterize the product (properties, costs, regulations, specifications, etc.), in order to identify the chemical reactions that would be appropriate for formulating it. In this sense, four chemical reactions were identified in section 2.1.3. The properties of the raw materials and the operating conditions of each chemical reaction are then defined in order to allow them to be evaluated. These characteristics are the initial data necessary for the calculation of the indicators (sub-criteria) characterizing the evaluation criteria of the decision model.

2.2.2.2. *Evaluation of decision criteria*

The second step consists of evaluating the chemical reactions under consideration through the defined set of criteria. For this purpose, it would be useful to return to the three-level hierarchical decision model detailed in Figure 2.1.

The four proposed evaluation criteria are represented by different sub-criteria. Each sub-criterion is defined by a mathematical equation allowing it to be evaluated as a function of the properties and operating conditions of the chemical reaction under consideration (temperature, cost of raw materials, time, pressure, etc.). Each of these equations makes it possible to obtain an index between 0 and 1, representing a certain form of "criticality". Thus, values close to 0 are the "best" values for this sub-criterion, and the values close to 1 are considered to be poor.

The sub-criteria are then aggregated through an average calculation to reach a value between 0 and 1 for each of the four criteria being considered. Thus, this step was carried out for reactions 1, 2, 3 and 4, which allow for ethylacetate production.

The results obtained are presented in Table 2.4.

	Economic	Environment	Safety	Occupational health
Reaction 2.1	0.526	0.507	0.156	0.521
Reaction 2.2	0.571	0.345	0.351	0.471
Reaction 2.3	0.541	0.63	0.263	0.521
Reaction 2.4	0.437	0.564	0.483	0.579

Table 2.4. *Evaluations of decision criteria*

This table highlights the strengths and weaknesses of each reaction, and at first glance, there does not seem to be an ideal reaction. For example, reaction 4 is very interesting from an economic point of view (with a value closest to 0), but it shows the worst results in terms of the other three criteria (environment, safety and occupational health). Reaction 2, on the other hand, seems very relevant from an environmental and health point of view, but it is significantly less advantageous from an economic and safety point of view.

So, from this table, it would seem difficult to make a relevant decision. It is necessary to find a compromise, and most importantly, to more accurately define the decision model. To this end, a reflection on the characterization of the decision criteria can be considered: can we consider that all evaluation criteria are on an equal footing? What are the priorities of the decision maker(s): economic profitability? The safety of the operations? The health of the staff, or the impact on the environment? The answer to this question very much depends on the context in which the decision is made and will be reflected in the weighting of evaluation criteria. What about in the particular context of the chemical industry, or even more precisely, that of the formulation of ethylacetate?

2.2.2.3. *Weighting of the decision criteria*

The weighting of the criteria for the model is a way to obtain a more realistic view and offers more contextualized decision support. This step is based on the pairwise comparisons made with the Saaty scale, presented in the previous section (Figure 2.2).

Two groups of experts have been asked to carry out this weighting. The first group consists of 23 practitioners from the chemical industry (plastic, petroleum, sales, academics and researchers). The second group consists of 22 students in the final stages of studying chemical engineering. A survey was conducted among these two groups of experts in order to determine the relative importance of the four criteria defined in the decision model. Each of them was asked to fill out a questionnaire that made it possible to realize a pairwise comparison of the four criteria: economy, environment, safety, and occupational health using the Saaty scales (Figure 2.5).

The consistency of the answers provided was then evaluated, in order to retain only the answers considered as eligible. Traditionally, the threshold for the consistency index (CI) of the AHP method is set at 10%, but in view of the complexity of the problem, a threshold below 20% was considered acceptable. Therefore, group 1 of the experts was reduced to 20 responses, and group 2 was reduced to 13 responses. These data thus made it possible to calculate a relative average weight by criterion, presented in Figure 2.6.

Select a number from 1 to 9 that best indicates the comparative importance of each criterion. Note that the importance chosen is always comparative between two criteria.

The assessment scale is shown in Figure 2:

Figure 2.

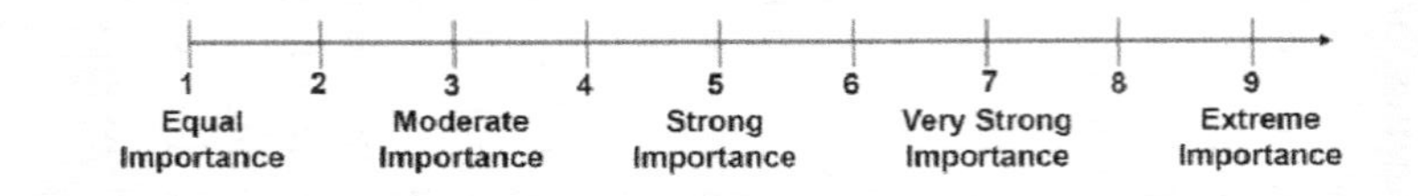

For example, if, in your view, the economic dimension is "strongly" more important than the environmental dimension, you should pick "5" on the left-hand side.

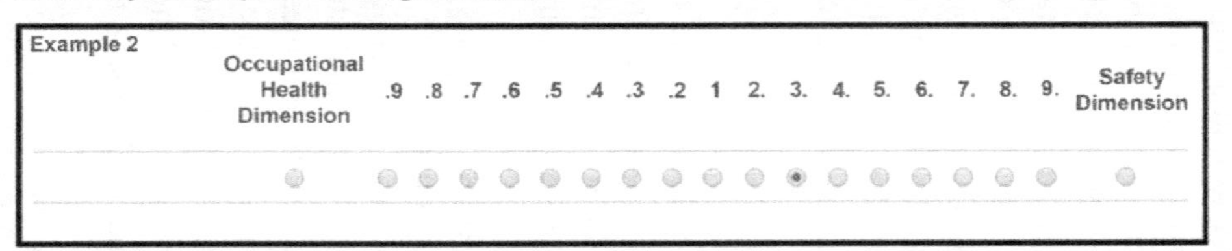

Figure 2.5. *Excerpt from the questionnaire of Serna et al. (2016)*

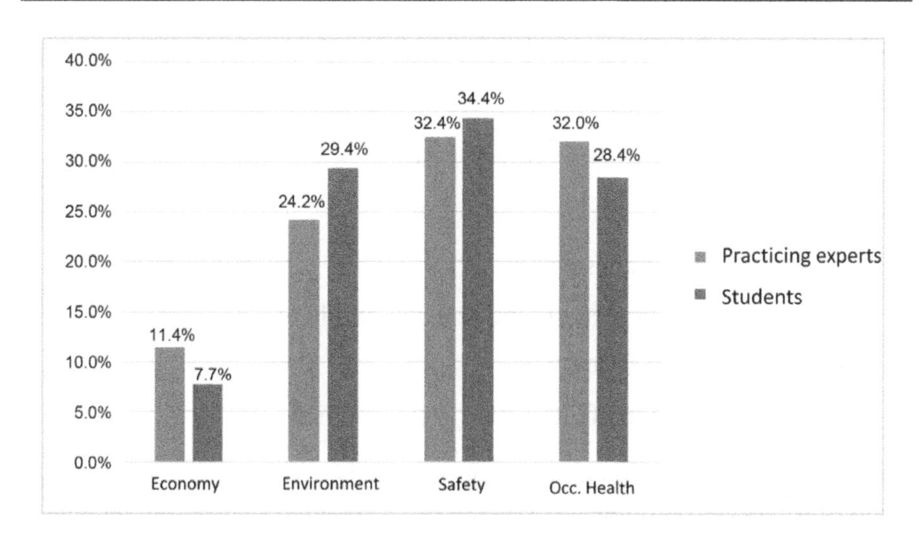

Figure 2.6. *Weighting of the criteria based on the two groups of experts. For a color version of this figure, see www.iste.co.uk/enjolras/decisionmaking.zip*

It is interesting to note that the weightings of the results obtained for the two groups are relatively similar. Safety comes first, followed closely by health and environment. Finally, economy is ranked last. These results reflect a certain awareness of the importance of social and environmental aspects in the chemical industry, both among students and practicing experts. However, there is another possible explanation for these results. Intuitively, when responding to this type of comparison, practicing experts take into account existing interrelationships between the criteria. For example, a safe and reliable process, by its nature, will be less expensive than a less secure process, which will require additional investments to control its potential risks. The reliable process will effectively involve fewer operational costs during its operation, which has a positive effect on the economic criterion. Therefore, in considering the safety criterion to be the most important, the practitioner intuitively indicates an impact on the economic dimension. The weight of the economic criterion is therefore relatively low, not because it is not important, but because the other criteria will favorably impact it.

2.2.2.4. *Decision*

The last step is to apply the weighting obtained to the evaluations of each reaction being considered. The AHP method proposes an aggregation by means of a weighted average, making it possible to obtain a single score in order to translate the performance of an alternative and compare it with the scores of others:

$$\mathit{Alternative}\ \text{score}\ A = \sum_{i=1}^{n} \omega_i \cdot E_{A,i},$$

where:

- A is the alternative considered;
- n is the number of criteria in the decision problem;
- ω_i is the weight of criterion i;
- $E_{A,i}$ is the evaluation of alternative A for criterion i.

However, it is important to carefully consider the way in which the criteria for carrying out this aggregation are evaluated. Indeed, the intended objective is particularly important for this calculation: are we trying to minimize or maximize the criterion?

In this case, we aim to have the lowest possible rating (see section 2.2.2.2). But the aggregation is based on a calculation of a weighted sum, or mathematically speaking, the highest and best value. There is therefore an inconsistency to be removed for the calculation of aggregate scores. The desired outcome is that the mathematically weakest evaluation should in fact be considered the best.

One of the options is therefore to aggregate the inverse value of each evaluation. Thus, the highest value becomes the lowest, and vice versa. Another possible option, in a case like ours where all the criteria are to be minimized, is to consider that the aggregate score itself is to be minimized. Thus, the alternative with the lowest score is the best.

We have chosen to calculate the aggregate scores according to the first option, with the input of the evaluation values. So, we apply the following calculation:

$$\mathit{Alternative\ score}\ A = \sum_{i=1}^{n} \omega_i \cdot \frac{1}{E_{A,i}},$$

where:

- A is the alternative considered;

– n is the number of criteria in the decision problem;

– ω_i is the weight of criterion i;

– $E_{A,i}$ is the evaluation of alternative A for criterion i.

Thus, applying this formula to the values in Table 2.4, the aggregate scores of the reactions are calculated according to the point of view of the two types of actors involved: practicing experts and students (Table 2.5).

Weightings of the experts	11.4%	24.2%	32.4%	32.0%	Score of the practitioners	Score of the students
Weightings of the students	7.7%	29.4%	34.4%	28.4%		
	Economic	Environment	Safety	Occup. health		
Reaction 2.1	1.901	1.972	6.410	1.919	3.385	3.477
Reaction 2.2	1.751	2.899	2.849	2.123	2.504	2.570
Reaction 2.3	1.848	1.587	3.802	1.919	2.441	2.462
Reaction 2.4	2.288	1.773	2.070	1.727	1.913	1.900

Table 2.5. *Ranking of alternatives according to the two expert groups*

Thus, we can observe, on the one hand, that the reaction scores are relatively close. It does not appear that one of the alternatives actually stands out from the group. On the other hand, we can note that the two groups of experts classify the reactions in the same order. Indeed, since the weightings of the groups of experts are relatively close to each other, this has little effect on the final ranking. Reaction 1 seems to be the preferred alternative, having earned the highest score for both groups of participants. It is closely followed by reactions 2 and 3, whose scores are almost equal, despite evaluations that are nevertheless well differentiated according to the criteria. Finally, reaction 4 scores last.

However, there are two potential limitations to these results. First, the mathematical calculation of AHP of the aggregate score of the alternatives is based on a weighted sum. The main limitation of this computational method is that it is compensatory. Indeed, if an alternative is very poorly evaluated for one of the criteria taken into account, despite the weighting associated with the criteria, this weak point will be compensated by the other evaluations, especially if they are very good. This is the principle of the end-of-semester average for students. Let us say a student receives an average grade in four subjects, a very good grade in one, and a very bad grade in the last subject. The student's weak point will be compensated for (in part) by the good grade, according to the coefficients (weights) assigned to each subject. Depending on the context, this phenomenon of compensation may or may not be problematic. In the case of the student at the end of the semester, this does not seem to be a problem. On the other hand, let us now imagine the evaluation of a food product during a consumer test. Can the criterion associated with the taste of the food be compensated for by its aesthetics or texture? It does not seem to make any sense. This compensation phenomenon would also seem unacceptable in the case of criteria relating to the concepts of safety. Thus, the phenomenon of compensation must be taken into account, and may prove to be limiting, or even to be prohibitive, depending on the context of the decision.

Another limitation of the AHP methodology lies in the fact that it considers the criteria to be independent of each other. However, it is very rare in a decision-making problem to build a framework of criteria that have no linkages or influences between them. This is a simplification of reality that, depending on the context, may or may not prove problematic. In our case, we consider that the interdependence between the criteria was an interesting point to expand upon. For this, AHP can be combined with another method: decision-making trial and evaluation laboratory (Dematel).

2.3. Going further: the question of interdependence between criteria

Innovation issues are inherently complex issues. That is why it is essential to consider them through what is known as a systemic vision. This approach is based on the notion of a system which, according to Joël de Rosnay (1975), can be defined by "a set of elements in dynamic interrelation, organized according to a purpose". Thus, if one wishes to come closer to reality and propose a more precise analysis, the elements of a decision problem in an innovation context cannot be considered independent. This is the main interest of the Dematel method. If AHP makes it possible to compare the criteria in terms of importance, Dematel reasons in terms of influence.

2.3.1. *The Dematel method*

The Dematel method was also created in the 1970s (Fontela and Gabus 1976). Its main interest is to make it possible to evaluate and quantify the influence exerted by different elements among themselves. Thus, in the case of a decision problem, Dematel enables the modelization of the influences that are received and exerted by the criteria considered by the decision maker.

It operates similarly to AHP. In fact, Dematel is also based on a principle of pairwise comparison. However, while the comparison in AHP is carried out through considering criteria of importance, with Dematel, this is done in terms of influence. Comparisons are made by experts who are able to identify the relationships between the criteria and quantify them. In some cases, decision makers can carry out this task themselves, and in other cases it is necessary to use an external expertise.

Let us take the example of buying a car using the three criteria: price, power and comfort. The pairwise comparisons in the Dematel framework will then take the form shown in Figure 2.7.

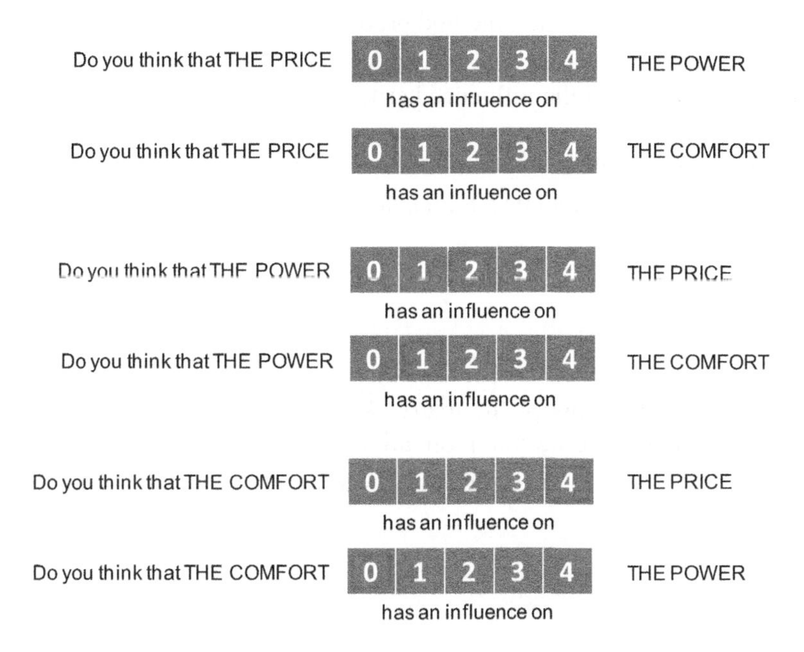

Figure 2.7. *Example of comparison of influences between criteria. For a color version of this figure, see www.iste.co.uk/enjolras/decisionmaking.zip*

This time, the scales will go from 0 to 4. A value of 0 means that criterion A does not exert any influence on criterion B. A value of 4 means that criterion A exerts a very strong influence on criterion B. It is interesting to note that the comparisons are carried out in both directions (from A to B and from B to A) because one criterion can influence another, without being influenced in the same way by that criterion. The results are thus grouped in an influence matrix $\vec{D}$, where the values along the diagonal are represented by zeroes (as a criterion cannot influence itself).

	Price	Power	Comfort
Price	0	0	0
Power	2	0	1
Comfort	4	0	0

Table 2.6. *Influence matrix* $\vec{D}$

The results of the Dematel method produce an influence map which, from a matrix calculation based on this matrix $\vec{D}$, allows us to visualize the influencing criteria and the influenced criteria.

As a result of a normalization of matrix $\vec{D}$, the matrix $\vec{B}$ is obtained, thus making it possible to calculate a total *influence matrix* named $\vec{X}$:

$$\vec{X} = \vec{B} \cdot (\vec{I} - \vec{B})^{-1},$$

with $\vec{I}$ the identity matrix.

The sums of each row of the matrix $\vec{X}$ are called r_i and define the total influence exerted by criterion i on all other criteria. The sums of each column of this matrix $\vec{X}$ are called c_i and define the influence of all criteria on criterion i.

The influence map (Figure 2.8) is thus constructed from the values of r_i and c_i. The x-axis represents the values of $r_i + c_i$ and the y-axis represents the values of $r_i - c_i$. In the case of buying a car, the influence map can be given as follows.

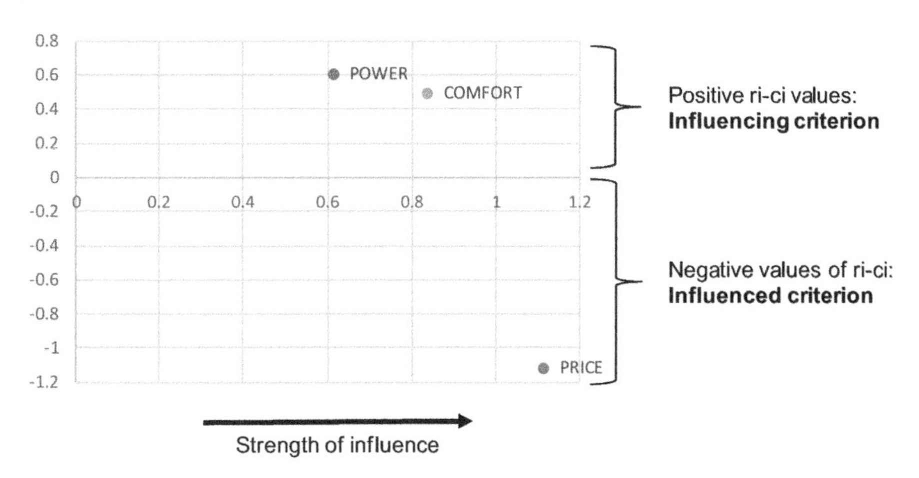

Figure 2.8. *Influence map for buying a car*

Based on the position of the criteria, this influence map (Figure 2.8) makes it possible to identify the strength and influence direction of these criteria. In this way, the criteria located on the upper part of the map (positive r_i-c_i values) are the influencing criteria. These are the values that generally exert more influence than they receive (in this case, comfort and power). The influenced criteria are located on the lower part of the map (in this case, the price). Finally, the position of the criteria on the horizontal axis reflects the strength of their interrelation with the decision-making system, or in other words, the strength of their contribution in the decision-making process. In our case, power and comfort are both influencing criteria, but it would seem that comfort has a stronger interrelation than power in the decision problem (it is located further to the right of the map). In the same way, the strength of the interrelation of the criterion of price seems particularly noticeable.

Part II. Criteria Influence Analysis

Select a number from 0 to 4 to indicate the influence of dimension A on dimension B, according to the following scale:

0	Without influence
1	Weak influence
2	Middle influence
3	Strong influence
4	Very strong influence

Dimension A			on			Dimension B
Economic Dimension	0	1	2	3	4	Environmental Dimension
○	○	○	○	○	○	○
Economic Dimension	0	1	2	3	4	Safety Dimension
○	○	○	○	○	○	○

Figure 2.9. *Excerpt from the Dematel of Serna et al. (2016)*

2.3.2. *Application to the case study*

In order to complete the results of our application case through the AHP method, a new questionnaire was submitted to the two groups of experts (practitioners and students). This made it possible to evaluate the influence between the criteria of the decision problem, following the principle of pairwise comparisons.

The results obtained are shown in Figures 2.10 and 2.11. It is interesting to note that if the criteria seemed to have weights equivalents in terms of importance for both groups of experts (Figure 2.6), in this case their opinions differ for the influences exerted and received.

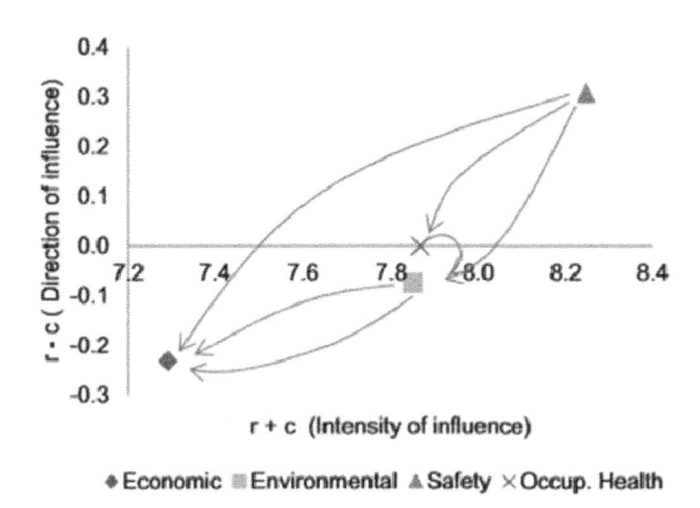

Figure 2.10. *Influence map of the expert group 1: practitioners*

Expert group 1 (practitioners: industrialists and researchers) considers that safety has a strong influence on all the other criteria, and that the economic criterion is impacted by all the other criteria. According to the influence diagram (Figure 2.10), the only criterion categorized as causal/influencing (as defined by Dematel) is safety. The others are affected/influenced criteria.

On the other hand, expert group 2 (chemical engineering students) considers that the criterion of the environment is the one that influences all the others, including safety. This time, there are two causal criteria: safety and environment, and two affected criteria: economic and occupational health criteria.

So, while for the practitioners (group 1), safety has the strongest influence, for the students, it is the environmental criterion that is of the greatest importance. This then reflects a different perspective on the issue of the environment, seen as either an end in itself (for practicing experts) or as a means (for students).

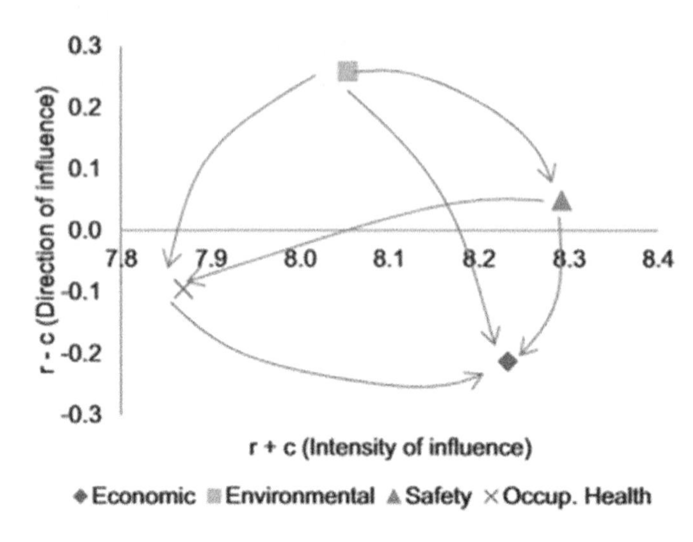

Figure 2.11. *Influence map from expert group 2: students*

2.3.3. *Proposal for an AHP/Dematel combined approach*

Taken separately, the AHP and Dematel methods each offer an interesting view of the decision problem and its characteristics. But the real interest lies in the combination of these two approaches: importance and influence. It may be the case that one criterion is considered extremely important for the decision, but then turns out to be strongly influenced by the other criteria considered. This is usually the case for economic criteria. Their importance is critical, but they are also affected criteria. In the future, one criterion may be considered unimportant in the decision and yet have a strong influence on the other criteria considered. This is often the case with environmental criteria, which in certain contexts may appear as secondary, but which will nevertheless have a direct impact on costs and technical, logistical and organizational methods, etc. Thus, to make a decision in the most informed way possible, it is useful to combine these two approaches in order to build a model that simultaneously grants importance and influence, to reflect reality as closely as possible.

Serna et al. (2016) have thus proposed an index combining AHP and Dematel in order to propose hybrid weights, taking into account both the importance of a criterion and its influence exercised and received: the cumulative sustainability index.

This index assumes that an affecting criterion should have a weight that is slightly larger than that which was originally calculated by AHP, since a percentage of its importance is distributed to the other criteria. In the same way, an affected criterion should have a slightly lower weight than that calculated by AHP, since a percentage of its importance is due to the influences received. Thus, Dematel acts as a corrective in the form of a reward/penalty scheme that allows the weightings calculated through AHP to be rectified in order to take into account the interrelation between the criteria.

Thus, for each criterion, an integrating weight is calculated. The related mathematical formula varies depending on whether the criterion under consideration is influencing (rewarded) or influenced (penalized) in the sense of Dematel.

Integrative weight of an influencing criterion *i*:

$$\omega_{integrative,i} = \omega_i + P \cdot \frac{r_i + c_i}{\sum_{affecting\ criterion} r_j + c_j},$$

where:

– ω_i is the weight of the calculated criterion *i* through AHP;

– P is a term between 0 and 1, allowing to regulate the consideration of Dematel;

– $\sum_{j\,:\,\text{affecting criterion}} r_j + c_j$ is the sum of the values of $r_j + c_j$ for all influencing criteria of the decision problem.

In the same way, the integrative weight of an influenced criterion *i*:

$$\omega_{integrative,i} = \omega_i - P \cdot \frac{r_i + c_i}{\sum_{j:\text{affected criterion}} r_j + c_j},$$

where:

– ω_i is the weight of the calculated criterion i through AHP;

– P is a term between 0 and 1, allowing to regulate the consideration of Dematel;

– $\sum_{j\,:\,\text{affected criteria}} r_j + c_j$ is the sum of the values of $r_j + c_j$ for all influenced criteria of the decision problem.

The definition of the term P amounts to choosing the part of consideration that we wish to give to Dematel in calculating the integrating weight. Would it be desirable for the influence between the criteria be taken into account to the same extent as the importance that it represents for decision makers (in this case, P = 50%)? Do we want to give preference to one over the other? Traditionally, the influence between criteria is considered but on a smaller scale than the importance of each of them. In fact, it is important to ensure that the value of the term P does not exceed the value of the weighting of the lowest of all the criteria, at the risk of seeing negative integrative weights appear. Influence plays a corrective role within the model. The value of P is generally relatively low (less than 50%).

Finally, the integrative sustainability index of an alternative is aggregated by a weighted sum calculation equivalent to that proposed in section 2.2.2.4:

$$Integrative\ sustainability\ index\ A = \sum_{i=1}^{n} \omega_{integrative\ ,i} \cdot \frac{1}{E_{A,i}},$$

where:

– A is the alternative concerned; i is the criterion concerned;

– n is the number of criteria in the decision problem;

– $\omega_{integrative,i}$ is the integrative weight of criterion i;

– $E_{A,i}$ is the evaluation of alternative A for criterion i.

Thus, the calculation of the integrative sustainability index for P-values between 0 and 15% gives the results shown in Figures 2.12(A) (practicing experts) and 2.12(B) (students). For both groups of experts, it is interesting to note that considering the influence between criteria makes it possible to widen the gap between alternatives. Effectively, the higher the P-values are, the more dispersed the ranking will be. However, in both groups, reaction 1 remains the best. However, we can visualize a change in the ranking for the group of practicing experts. Reaction 2 is initially ranked second, and for $P > 5\%$, it moves to third. This is a minimal change, but it nevertheless reflects the correction applied by taking into account the influences between criteria.

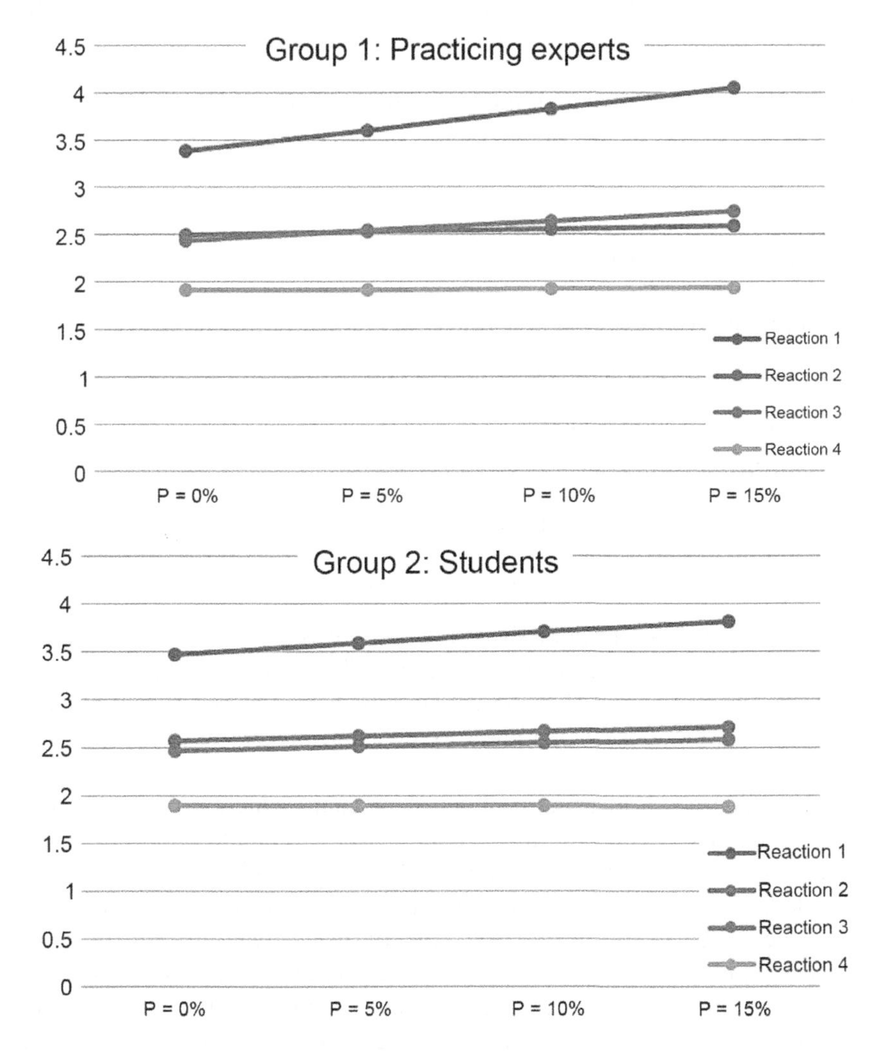

Figure 2.12. *(A) Change of the ranking of reactions according to the cumulative sustainability index – practicing experts; (B) evolution of the classification of reactions according to the cumulative sustainability index – students. For a color version of this figure, see www.iste.co.uk/enjolras/decisionmaking.zip*

2.4. AHP: instructions for use

Applying the AHP method requires following different steps. Although these are based on notions of linear algebra, the related mathematical calculations are relatively simple and allow for the method to be applied using a classic spreadsheet. However, there are many free software programs that make it possible to use AHP in a more intuitive way.

In this section, we present a practical guide that gives the various steps of the method as well as an application example using the Total Decision software.

2.4.1. *Practical guide*

To apply the AHP method, this practical guide offers four steps to follow:

– the hierarchical structuring of the decision problem;

– the weighting of the criteria;

– identification and evaluation of alternatives;

– aggregation of results and decision.

2.4.1.1. *The hierarchical structuring of the decision problem*

As with all multi-criteria analysis methods, it is first necessary to define the elements to be considered in the decision. Thus, for AHP, a hierarchical repository must be built (Figure 2.13). Level 1 contains the evaluation criteria, that is, the elements studied by decision makers to determine their choices. In the case of buying a car, these criteria can be price, comfort, power, etc. There is a maximum of seven of them, which are considered to make the method easier to implement. Level 2 contains sub-criteria, if needed, to better characterize the criteria of Level 1 once they are aggregated. For example, the sub-criteria of the criterion "price" of the car can be the purchase price and the cost of use (including consumption, maintenance, insurance, etc.). Once these two sub-criteria are aggregated, we get the "total price" of the car. Finally, level 3 contains the indicators for evaluating the criteria and sub-criteria. For example, the power of the car is assessed by the horsepower rating of the engine. Comfort is assessed by the volume of the passenger area (in m^3), etc.

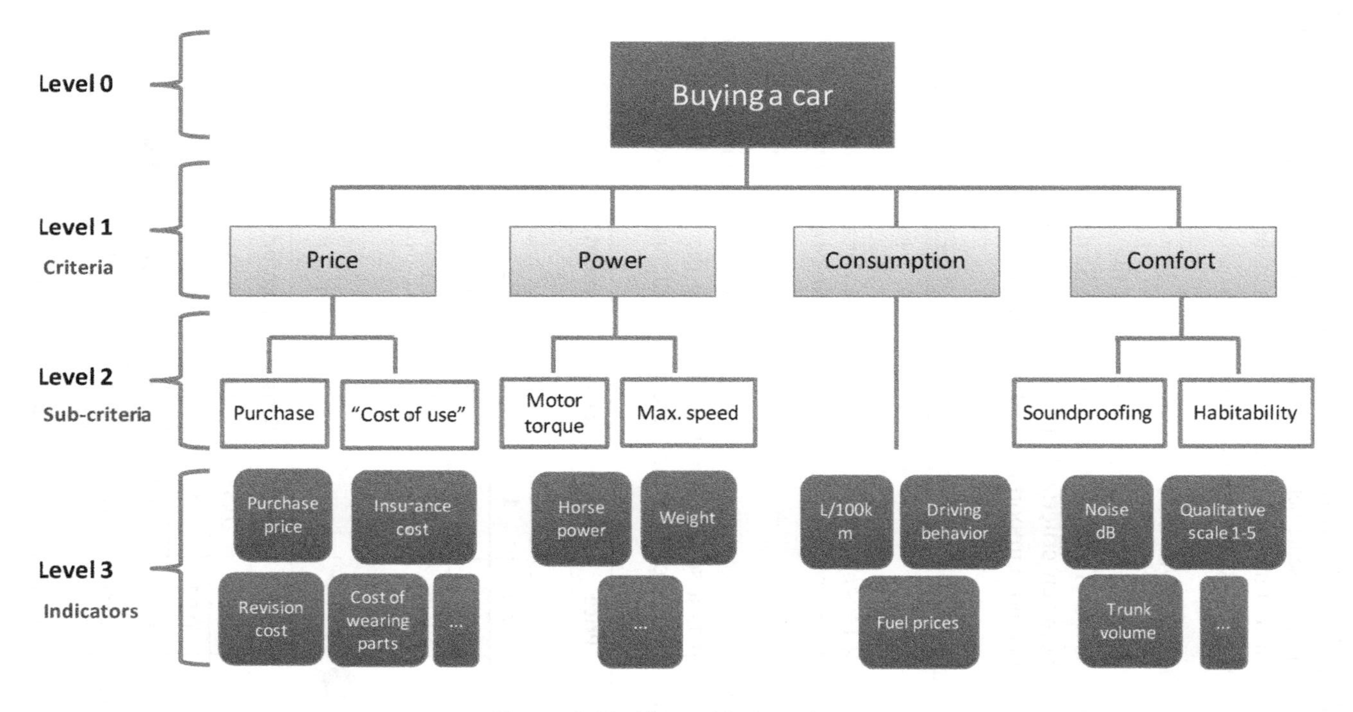

Figure 2.13. *Hierarchical model*

CAUTIONARY NOTE 1.– To evaluate more "subjective" criteria, indicators can also be given on a numerical scale between 1 and 10, such as for expressing a ranking without defined units of measurement. The aesthetics of the structure, for example, could also be assessed in this way by decision makers.

2.4.1.2. *Weighting of the criteria*

The protocol for calculating the weighting of the criteria begins with the pairwise comparisons of the criteria, through the Saaty scales. In the case of the comparison between criteria A and B, the decision maker must identify which of the two is more important, and how much so.

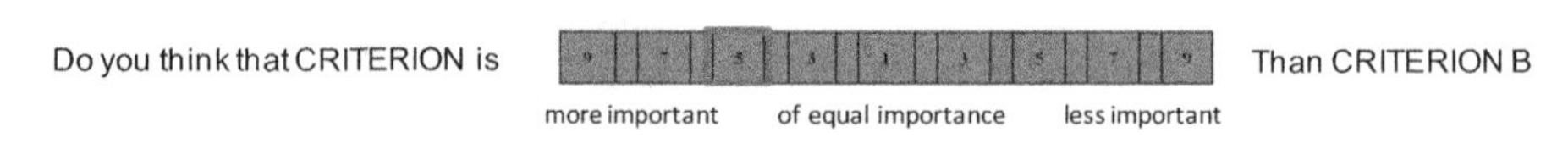

Figure 2.14. *Application of the Saaty scale. For a color version of this figure, see www.iste.co.uk/enjolras/decisionmaking.zip*

By combining the set of pairwise comparisons, the comparison matrix (A) is obtained. It is a square matrix comprising the value 1 over its entire diagonal. Then, the values on either side of the diagonal are reversed.

2

	Price	Power	Consum.	Comfort
Price	**1**	3	1/2	5
Power	1/3	**1**	1/5	2
Consum.	2	5	**1**	7
Comfort	1/5	1/2	1/7	**1**
Sum	3.53	9.5	1.84	15

3 *Normalization*

Level 1	Price	Power	Consum.	Comfort
Price	0.28	0.32	0.27	0.33
Power	0.09	0.11	0.11	0.13
Consum.	0.57	0.53	0.54	0.47
Comfort	0.06	0.05	0.08	0.07

4 *Average*

Weight
0.30
0.11
0.53
0.06

Figure 2.15. *Illustrated approach: calculation of the weights of the criteria*

This matrix is then normalized to obtain the matrix A′. A common method of standardization involves dividing the initial values of the matrix A by the sum of the corresponding column (distributive normalization). However, other methods can also be considered (see section 6.2.2.3).

Next, the vector of the weights ω is calculated by averaging each row of the normalized matrix (A′) to obtain the weights of each criterion (Figure 2.15).

CAUTIONARY NOTE 2.– The sum of the weights must always be equal to 1.

Finally, in order to validate the weights obtained, the consistency ratio RC is then calculated. According to Saaty, if the RC ratio is less than 10%, the consistency is judged to be good. However, if the number of pairwise comparisons to be performed is large, a higher tolerance (20%, for example) is acceptable.

The first step is to calculate λmax, which represents the average of the eigenvalues of the matrix (A). Then, the value retained for λmax is equal to the average of the values of the eigenvector of (A).

$$Eigenvector\ (A) = \frac{\vec{A} \cdot \vec{\omega}}{\vec{\omega}}$$

CAUTIONARY NOTE 3.– This calculation requires the creation of a matrix product $\vec{A} \cdot \vec{\omega}$, which makes it possible to obtain a vector. This vector can then be divided term by term by the vector $\vec{\omega}$.

Next, the consistency index IC must be calculated:

$$IC = (\lambda max - K)/(K-1),$$

where K is the number of criteria.

Finally, the consistency ratio RC is obtained by dividing IC by the random index of Saaty IA.

$$RC = IC/IA.$$

The random index IA is a standard value that depends on the number of criteria that can be found in Figure 2.3.

2.4.1.3. *Identification and evaluation of alternatives*

Once the set of criteria has been determined and weighted, a finite number of alternatives to compare must be selected. The decision will therefore consist of identifying, from among these alternatives, which one(s) is (are) the best, according to the preferences of the decision maker. The AHP method makes it possible to compare a large number of alternatives with each other. However, each one must be evaluated according to the defined criterion reference base (Figure 2.16).

CAUTIONARY NOTE 4.– In this stage of evaluation, it is crucial to pay attention to the nature of the criteria and the objectives associated with them. In fact, some criteria are to be maximized (such as comfort) and others are to be minimized (price). These objectives must be kept in mind when evaluating alternatives, in order to maintain mathematical consistency. As an example, Table 2.14 evaluates the price of the car in thousands of euros. The best car is therefore Car X, which has the lowest price value. However, from a purely mathematical point of view, it is the highest value that will positively impact the final score. There is thus a mathematical inconsistency between the mathematical significance of the price value and its decision-making objective. This case can easily be solved by inverting the scale of values for this criterion: instead of evaluating the cars according to their price, they will be evaluated according to the inverse of their price. Thus, the cheapest car gets the best rating, and consistency is restored.

CAUTIONARY NOTE 5.– The evaluation scales. The aggregation method of AHP is based on a weighted sum calculation. This method of calculation is compensatory. This means that if we aggregate elements with very different orders of magnitude, the larger orders of magnitude will compensate for the smaller ones. The next order would then be to normalize the matrix of evaluations, in order to reduce this compensation effect.

1 *Identification of objectives*

	Minimizing	Maximizing	Minimizing	Maximizing
Cars	Price (k€)	Power (HP)	Cons. (L/100km)	Comfort (1-5)
Medium	18	75	8	3
Sport	35	110	9	2
Luxury 1	40	90	8.5	5
Economic	8.5	50	7.5	1
Luxury 2	38	85	9	4

	Minimizing	Maximizing	Minimizing	Maximizing
Cars	Price (k€)	Power (HP)	Cons. (L/100km)	Comfort (1-5)
Medium	1/18	75	1/8	3
Sport	1/35	110	1/9	2
Luxury 1	1/40	90	1/8.5	5
Economic	1/8.5	50	1/7.5	1
Luxury 2	1/38	85	1/9	4

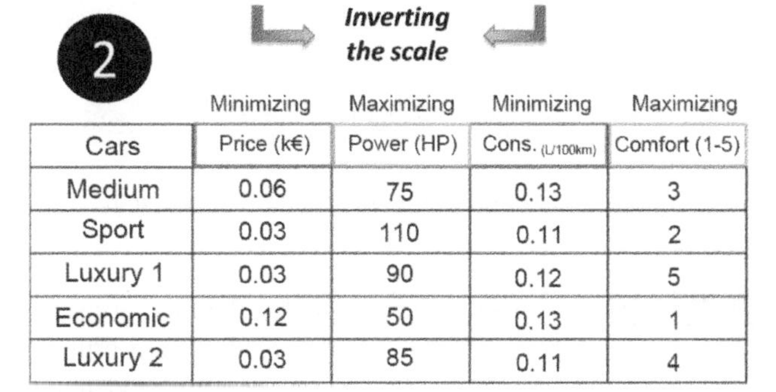

	Minimizing	Maximizing	Minimizing	Maximizing
Cars	Price (k€)	Power (HP)	Cons. (L/100km)	Comfort (1-5)
Medium	0.06	75	0.13	3
Sport	0.03	110	0.11	2
Luxury 1	0.03	90	0.12	5
Economic	0.12	50	0.13	1
Luxury 2	0.03	85	0.11	4

Normalization

	Minimizing	Maximizing	Minimizing	Maximizing
Cars	Price (k€)	Power (HP)	Cons. (L/100km)	Comfort (1-5)
Medium	0.25	0.23	0.27	0.27
Sport	0.125	0.34	0.22	0.18
Luxury 1	0.125	0.28	0.24	0.45
Economic	0.5	0.15	0.27	0.09
Luxury 2	0.20	0.24	0.25	0.35

Figure 2.16. *Illustrated approach: evaluation of alternatives*

2.4.1.4. *Aggregation of results and decision*

Finally, the final step is to obtain a ranking of the different alternatives by aggregating their evaluations into a single score (Figure 2.17). Taking this into account, the AHP method is based on a weighted sum calculation, where for each alternative, its own evaluation values according to each criterion will be weighted, and then aggregated. The higher the score, the higher the alternative is in the ranking. This therefore gives an indication of how well it fits with the preferences of the decision maker.

4 *Calculation of scores*

	Minimizing	Maximizing	Minimizing	Maximizing	
Cars	Price (k€)	Power (HP)	Cons. (L/100km)	Comfort (1-5)	Score
Medium	0.25	0.23	0.27	0.27	0.26
Sport	0.125	0.34	0.22	0.18	0.20
Luxury 1	0.125	0.28	0.24	0.45	0.23
Economic	0.5	0.15	0.27	0.09	0.31
Luxury 2	0.20	0.24	0.25	0.35	0.22
Criteria weights	0.30	0.11	0.53	0.06	

Figure 2.17. *Aggregation of results – application*

CAUTIONARY NOTE 6.– It is important to clarify that the final decision belongs to the decision maker alone. Multi-criteria analysis methods are primarily tools to *support* decisions, and not “deciding tools”. They allow for modeling preferences in order to open the discussion, not to replace the decision maker.

2.4.2. *Illustration of related free software*

Several software programs have been developed to automate the application of the AHP method: Transparent Choice[1], Super Decision[2], Expert Choice[3], Total Decision[4], etc. These programs offer different

1 Available at: www.transparentchoice.com/ahp-software.

2 Available at: www.superdecisions.com/.

3 Available at: www.expertchoice.com/.

4 Available at: www.vilenio.com/td_download.html.

functionalities, ranging from the construction of hierarchical decision models to weighting from the criteria and even sensitivity analysis results.

We will focus on the example from Total Decision, software that is available for free and includes a large number of useful features, among others:

– creation of a hierarchical model;

– evaluation of alternatives;

– weighting decision criteria;

– aggregation and visualization of results;

– sensitivity analyses.

The use of Total Decision makes it possible to apply the method in a fast and intuitive way, while offering visual and interactive interfaces that facilitate exchanges with decision makers. In particular, for the weighting of the decision criteria, the Saaty scales are dynamic and make it possible to fill the comparison matrix A simultaneously, while calculating the related consistency ratio. In the case of a ratio that is too high, the software offers recommendations for how to improve it to reach an acceptable threshold. This step, applied to the example of buying a car, is illustrated in Figure 2.18.

The presentation of the results is also dynamic. It is possible to move the initial data such as the weights criteria in order to see how the ranking of alternatives is changing (Figure 2.19).

Finally, the sensitivity analysis allows for creating scenarios regarding decision-making. The graph that is obtained makes it possible to visualize the ranking of the alternatives according to the evolution of the weight of a given criterion. For example, Figure 2.20 shows the sensitivity analysis of the classification of cars as a function of the evolution of the criterion of comfort.

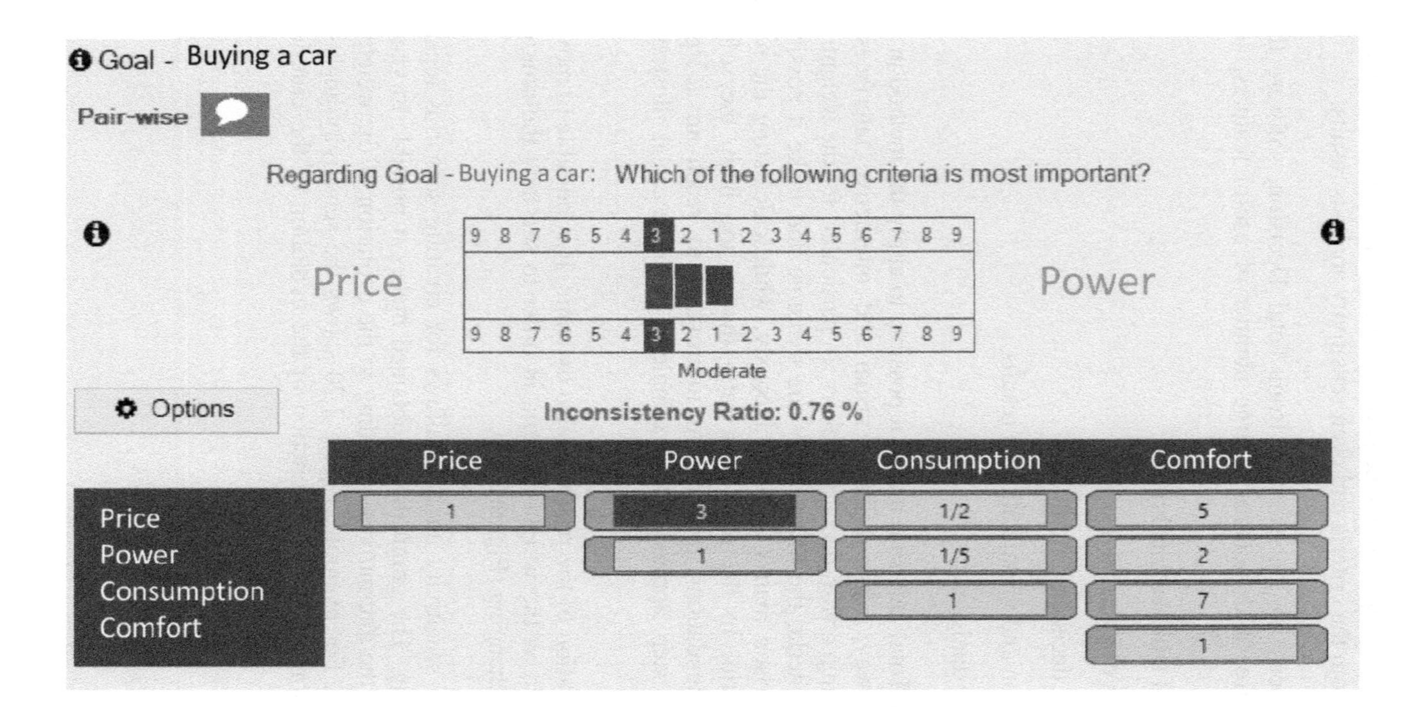

Figure 2.18. *Screenshot of the weighting phase criteria on Total Decision*

For low weighting values associated with comfort, the economy car comes out on top, but for a weighting value for comfort at 25%, we can see that there is a change in the ranking. The economy car falls behind the Luxury car 2, which now places first in the ranking. Thus, the value of 25% is a transition zone that marks a difference between two scenarios: with all other conditions equal, scenario 1 models the preferences of a decision maker who thinks comfort is important, and scenario 2 models the preferences of a decision maker who considers it to be secondary. This principle is equivalent to the principle used by Serna et al. (2016), which changes the consideration given to the Dematel method in the composition of the sustainability index (see section 2.3.3). These sensitivity analyses also enable finding compromises in decisions made among multiple stakeholders. They make it possible to find areas of agreement (where the ranking changes) and potentially identify areas for discussion, allowing two stakeholders to come to an agreement.

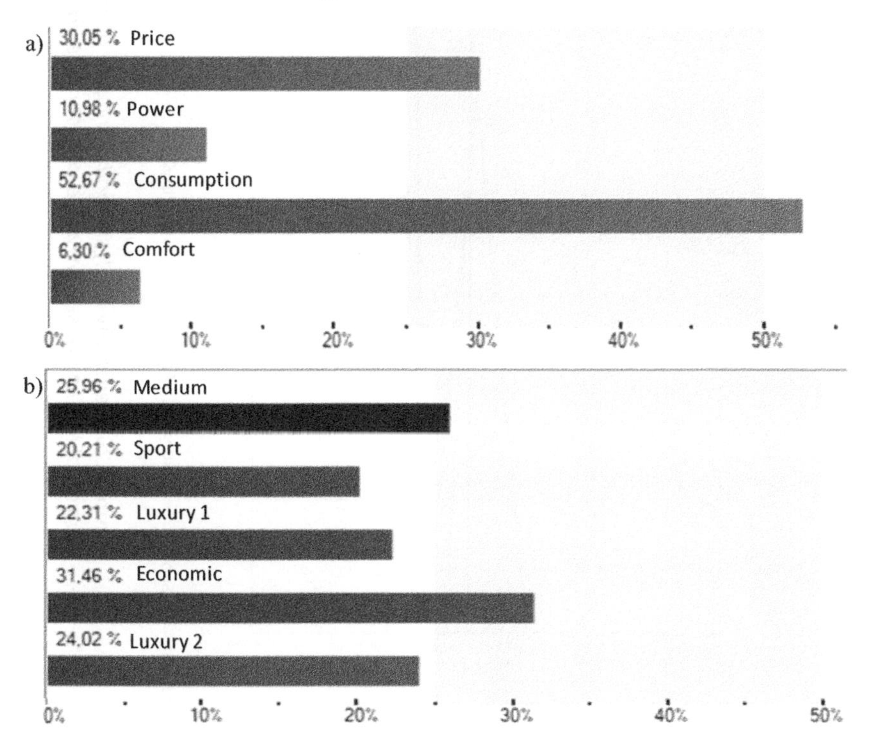

Figure 2.19. *(a) Screenshot of the weighting criterion and (b) the classification of alternatives*

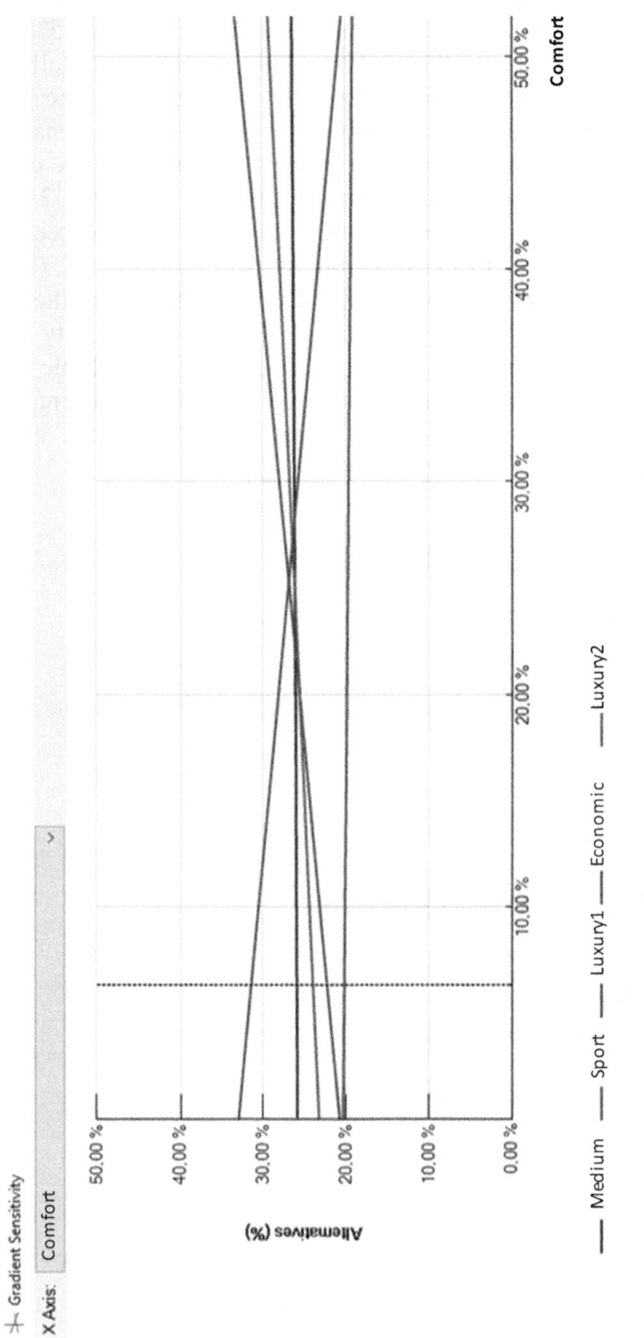

Figure 2.20. *Sensitivity analysis for the criterion of comfort. For a color version of this figure, see www.iste.co.uk/enjolras/decisionmaking.zip*

This means that using software that supports the AHP method allows for a more dynamic and interactive application, leading to results that are easier for decision makers to understand. Remember that the intended purpose of these decision-support methods is to put forward a model. The final decision always rests with the decision maker. Therefore, the use of software such as Total Decision then makes it possible to obtain a dynamic model, in order to better prepare and familiarize the people making the decision.

AHP AT A GLANCE.– Objective: AHP is a ranking method. Based on the calculation of an aggregated score, it seeks to classify alternatives from the best to the worst, according to the preferences of the decision-maker.

Its unique features are as follows: it includes a weight construction stage for the criteria through a pairwise comparison. It also makes it possible to check the consistency of these comparisons. These specific stages in AHP can be combined with other multi-criteria analysis methods to create hybridizations, and integrate the construction of weights into the decision problem.

Its limits are as follows: it is a compensatory method, since it is based on the mathematical principle of the weighted sum.

An additional note is that like many other methods, AHP considers the criteria to be independent. It can therefore be combined with the Dematel method, which makes it possible to discard this hypothesis and to consider the existing interrelations between the criteria.

2.5. References

Acosta, P., Acquier, A., Carbone, V., Delbard, O., Fabbri, J., Gitiaux, F., Manceau, D., Ronge, C. (2014). Les business models du développement durable. *L'Expansion Management Review*, 152, 20–29.

De Rosnay, J. (1975). *Le macroscope : vers une vision globale*. Points, Paris.

Fabbri, J. (2017). Fabriquer et concevoir l'innovation : des outils à portée de main. In *Management de l'innovation*, Gay, C. and Szostak, B. (eds). Dunod, Malakoff.

Fontela, G. (1976). The DEMATEL Observer. Battelle Institute Geneva. PhD thesis, Sunway University, Selangor.

Ghazinoory, S., Aliahmadi, A., Namdarzangeneh, S., Ghodsypour, S.H. (2007). Using AHP and L.P. for choosing the best alternatives based the gap analysis. *Appl. Math. Comput.*, 184, 316–321.

Hacking, T. and Guthrie, P. (2008). A framework for clarifying the meaning of Triple Bottom-Line, Integrated, and Sustainability Assessment. *Environ. Impact Assess. Rev.*, 28, 73–89.

Saaty, T.L. (1980). *The Analytic Hierarchy Process: Planning, Priority Setting, Resource Allocation*. McGraw-Hill, New York.

Saaty, T.L. (2008). Decision making with the analytic hierarchy process. *Int. J. Serv. Sci.*, 1, 83–98.

Serna, J., Díaz Martinez, E.N., Narváez Rincón, P.C., Camargo, M., Gálvez, D., Orjuela, Y. (2016). Multi-criteria decision analysis for the selection of sustainable chemical process routes during early design stages. *Chem. Eng. Res. Des.*, 113, 28–49.

3

Marketing Strategy During the Market Entry Phase: An Application of Rough Sets

The decision to innovate is a strategic commitment, requiring an organization to be able to look to the future methodically. Strategic decisions will arise throughout the innovation process in order to implement a successful market entry. This means that the final phase of the innovation process of a product/service, culminating in its launch on the market, must be planned out and prepared.

This includes a marketing approach that is well thought-out and structured. Yet, it is difficult to anticipate the impacts and effectiveness of this marketing approach. This in turn means the necessary efforts, both financial and temporal, are difficult to assess, and in view of the potential stakes, making decisions regarding the allocations of marketing resources must be done objectively and with the assistance of various tools to limit the risks inherent in launches of new innovations on the market.

To illustrate this problem, we will start on the basis of a study carried out by Mahapatra et al. (2010), aimed at identifying the patterns of the most relevant distribution of expenses for maximizing product sales in the cosmetics sector. This article provides an application of the Rough Sets method developed by Pawlak in the 1980s.

Decision-Making Tools to Support Innovation,
by Manon ENJOLRAS, Daniel GALVEZ and Mauricio CAMARGO.

3.1. Context and challenges in decision-making

3.1.1. *Decision-making in marketing*

For a new product or service, the transition from having the status of an invention to that of an innovation is determined by its introduction on the market. However, the marketing of a product or service is a critical phase, and the most frequently cited reasons for the failure of new products or services on the market are often related to the inadequate management of the marketing approach for such products (Kotler et al. 2012): misinterpretation of market research, excessively accelerated development, an overly delayed launch, an overestimated market size, poor promotion or unsuitable pricing policy, etc.

Faced with the high risks of failure, it becomes clear that it is important for a company to properly manage its marketing strategy in the development process in order to encourage customers to think about it and thus differentiate itself from its competitors. Generally speaking, marketing involves the management of a company's brands, the promotion of products, the retaining of customers and the creation of demand. However, this view is actually only a shortened version of the central role of marketing within the company's strategy. In fact, marketing has a strategic role that impacts the company as a whole, bringing together considerations on the profitability of products, commercial and technical monitoring, and the prospective vision of future planned developments, as well as changes in the market (Nagard-Assayag et al. 2015). Therefore, within the context of marketing-related issues, the decision-making process cannot be limited to a narrow vision of this field as a simple communication tool and sales booster. The existing interaction between marketing and other departments of the company (IT, production, development, etc.) must be brought to light. These interactions are particularly crucial when it comes to the development or R&D departments. In practice, all innovative design activities are not limited to a vision that merely covers technological feasibility, but instead are interconnected with an examination from a marketing standpoint, in which the customer is introduced to the rationale of the design (Pointet 2011).

However, this integrative view of marketing frequently leads to internal communication problems and forces teams to address contradictory issues. For example, the logistics/production department of a company will seek to limit the stock of finished products for operational and economic reasons that are obvious from their point of view. On the other hand, reduced inventory levels present a potential risk from the point of view of marketing and sales because this increases the risk of running out of stock, which leads to customer dissatisfaction. It is therefore necessary to identify these contradictory issues and integrate them into the decision-making process. The interest of incorporating multi-criteria decision support methods into this process would therefore make perfect sense.

Traditionally and historically, marketing has been characterized through the framework of the "4 Ps", established by J. McCarthy in 1964. While during the 1980s, this traditional model was reinterpreted many times to more closely reflect evolving consumption trends (Booms and Bitner 1980), these 4 Ps remain the foundation of the concept of a "marketing mix", and reflect the fundamental elements that define how a product offering is shaped. These then translate into the following operational areas:

– *product*: the choices relating to the characteristics of the product itself, but also to its packaging, design, normative constraints and product line policy;

– *price*: the strategic positioning of the company in terms of price policy, taking into account costs, competition and product image;

– *promotion*: the communication policy, product promotion and related direct marketing actions (e-mailing, prospecting, surveys, etc.);

– *place*: refers to the choice of strategy for the distribution of the product and the channels used to reach end customers.

However, the role of marketing varies considerably depending on the nature of the innovations. The importance assumed by each of the 4 Ps or their role in the process will therefore vary depending on the context. When innovation involves a shift in customer uses, market acceptance becomes a real issue, and marketing is required to play an essential role in "explaining" innovation (Nagard-Assayag et al. 2015). In this case, promotion becomes crucial. When innovation is based on a new business model, it is necessary to analyze in detail who is willing to pay, and the maximum amount they will pay. The definition of a relevant pricing policy, as well as the fine-level analysis of the value given to the proposal by each player in the ecosystem, is a priority in this case. Finally, when innovation is technological, it is necessary to be certain that the characteristics of the product itself are perceived and valued by potential customers. The valuation of the product therefore plays an essential role in this case (Adams et al. 2019).

Thus, implementing the marketing function and managing its interaction with the other activities of the company while remaining in line with the scope of the innovation project in progress involves many strategic decisions that must be anticipated.

3.1.2. *Definition of the decision model*

To illustrate these problems, we will use a case study conducted in the cosmetics sector. The article by Mahapatra et al. (2010) aims to identify the profiles of expenses which are the most relevant for maximizing sales of cosmetic products. This study is based on the analysis by a panel of 23 Indian companies in the sector, taken from the CMIE-Prowess database.

The data were collected over a 3-year period (2003–2005), and six parameters were selected to characterize these companies:

– *marketing – MKT*: this includes the amount of expenses incurred for the promotion of the company (organization of events, sales, direct marketing/e-mailing, etc.);

– *advertising – ADV*: it aggregates the expenses related to promotional announcements made through the media, such as television, radio, the Internet, newspapers;

– *distribution – DIS*: it relates to expenses for distributing the product to customers (logistics activities and value chain management);

– *R&D*: it relates to the research and development budget committed to developing the product;

– *miscellaneous – MIS*: it includes expenses related to corporate social responsibility, such as the launch and management of a foundation;

– *product sales made – PS*: it indicates the revenues obtained by the company over the period under consideration.

Thus, the first five parameters represent characteristics of companies, and the sixth parameter (product sales) represents the output data, which we are trying to improve (the higher the revenue, the more successful the company will be considered). All of the above parameters are valued in millions of Indian rupees. However, the researchers chose to switch from a quantitative measurement (in millions of Indian rupees) to a standardized qualitative measurement scale of "low/medium/high", formed on the basis of expert opinion. In this way, it is possible to propose inclusive categories:

– (Expenses/PS) = low: normalized value < 0.3;

– (Expenses/PS) = medium: 0.3 < normalized value < 0.7;

– (Expenses /PS) = high: normalized value > 0.7.

This decision-making model is therefore based on five criteria, evaluated according to a qualitative scale, and characterizing 23 companies that have obtained higher or lower revenues. The objective is therefore to identify the combination(s) of criteria to maximize the sales of the companies on the panel.

In the previous section, we mentioned the importance of considering marketing as an integrated activity that has an impact at different levels of the company. Since this case is related to a product-type problem, we chose to rely on the historical 4 Ps model to analyze the proposed decision model. In fact, orienting around an extended model would seem unnecessary here, since the context does not require the integration of additional elements.

It is interesting to note that this decision model mainly emphasizes the "promotion" element of the marketing mix. Indeed, the criteria of marketing (MKT), advertising (ADV) and miscellaneous (MIS) refer directly to the way the company communicates about the product and its image. Whether from an operational (MKT), communication (ADV) or strategic (MIS) point of view, this decision model highlights an important focus on this aspect.

However, the proposed decision model also incorporates the "product" element, through expenses in R&D, spent in order to help it evolve and adapt it to the needs of customers. It also incorporates the "place" element, through the distribution criterion (MIS), which translates the way the product is brought to the customer.

On the other hand, the "price" element of the marketing mix is not incorporated into this study. This may seem to be a missing element because the decision model does not represent the strategic position of the company.

That said, the choice to evaluate the criteria in the form of expenses may explain that this aspect is not integrated into the model because it is not properly adapted to the measurement scales in place.

3.2. The Rough Sets method or the theory of approximate sets

Developed in the early 1980s by the Polish mathematician and computer scientist Zdzisław Pawlak, the Rough Sets methodology is a descriptive multi-criteria analysis method.

Unlike most of the other methods presented in this book, its objective is not to classify or sort alternatives (a method of ranking or sorting) but to extract information from a data set in order to describe the implications and consequences of defined scenarios. This is a mathematical approach to dealing with imperfect knowledge, that is, the inaccuracy of a data set. The Rough Sets method makes it possible to extract hidden information from this data set and to generate approximations to reduce this uncertainty.

3.2.1. *Terminology*

The terminology used in the context of this method is based on the notion of an attribute. To define the problem to be solved, it is necessary to define *conditional attributes* and a *decision attribute*. Conditional attributes are the "measured" elements describing the data set. The decision attribute, on the other hand, conveys the outcome/intended purpose that we want to describe.

If we apply this terminology to our application case (see Table 3.1, next section), we then define:

– twenty-three examples: the Indian companies from the studied sample;

– five conditional attributes: (1) marketing expenses (MKT), (2) advertising expenses (ADV), (3) distribution expenses (DIS), (4) R&D expenses (R&D) and (5) miscellaneous expenditures (MIS);

– one decision attribute for the value of sales made by the company (PS).

Therefore, the Rough Sets methodology applied in this particular case will, on the one hand, build *subsets* (companies with low/medium/high sales) and will seek to identify the *rules* associated with these subsets.

These rules are defined in a conditional form: if "conditional attribute values", then "decision attribute value" or, more precisely:

> If "MKT = V, ADV = W, DIS = X, R&D = Y and MIS = Z" then "PS = low or medium or high",

with {V, W, X, Y, Z} = {low, medium, high}.

For example, one of the rules identified in our case could be translated in the following form:

> *if* the expenses of the company are *low* for marketing (MKT), *low* for advertising (ADV), *low* for distribution (DIS), *low* for R&D and *low* for miscellaneous elements (MIS), *then* the resulting revenue (PS) will be *low*.

Identifying these rules is the primary objective because these will identify the conditions to be implemented to improve the potential sales of the future product. On the basis of these rules, it is possible to estimate the minimum effort required for marketing to achieve a certain sales goal.

3.2.2. *Fundamental principle: indiscernibility*

The fundamental principle of the Rough Sets method is that of indiscernibility (Pawlak 1998). Indiscernibility is considered to be a case where certain examples are characterized by the same value of conditional attributes (though potentially different decision attribute values). These sets of indiscernible examples are then called *elementary sets*.

Indiscernibility is thus expressed through a boundary *region* from this set. This boundary region includes all elementary sets comprising different values of decision attributes.

Here, X will be a subset of examples, all of which have the same decision attribute value. To best describe this subset X, the Rough Sets method will allow us to define three different regions (Figure 3.1):

– a positive *region*: comprising the elementary sets being included in the set X in a certain way (and therefore having conditional attribute values systematically leading to a decision attribute value equal to X);

– a negative *region*: comprising the elementary sets outside of the set X in a certain way (and therefore where they all have a decision attribute value different from X);

– a boundary *region*: comprising the elementary sets that can either be included in the subset X, or be outside of it. The elementary sets of the boundary region therefore have identical conditional attribute values within the same set, but different ranges of decision attributes.

Thus, two approximations can be proposed in order to reduce the uncertainty associated with this boundary region (Sawicki and Żak 2014) (Figure 3.1).

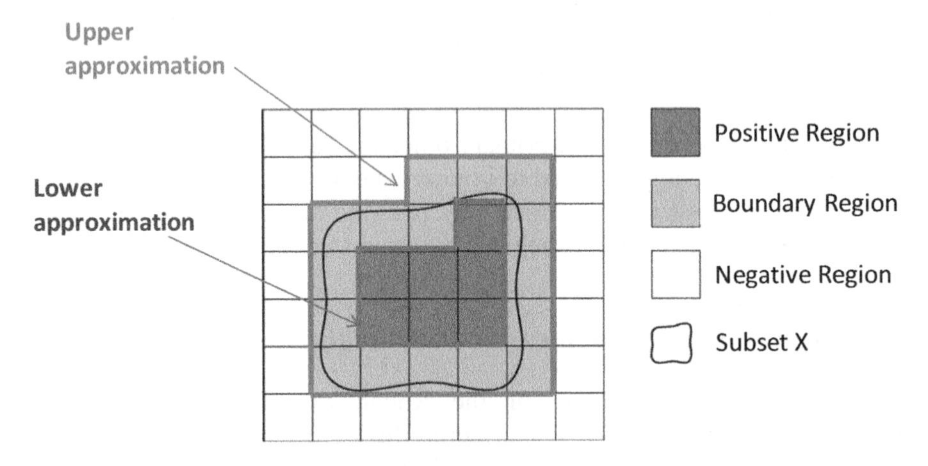

Figure 3.1. *Boundary region and approximations (source: Yao 2013)*

First, an upper approximation can be considered. This upper approximation consists of considering that the subset X includes the positive elementary sets of the region to which those of the boundary region are added. This upper approximation therefore represents *the set of examples that could potentially be included in X.* This is an optimistic approximation.

Second, a lower approximation may be preferred. This lower approximation considers that the subset X can be approximated by only the positive region. This approximation therefore represents *the set of examples included in X in a certain way*. It is a pessimistic approximation.

3.2.3. *Application of the method*

In order to clarify this notion of indiscernibility, we will apply this principle to our case study. Our study of 23 Indian cosmetic firms seeks to identify the combinations of necessary marketing expenses (conditional attributes) making it possible to arrive at a total sales value deemed acceptable (decision attribute). We can then address behavioral models that we will try to reproduce or avoid, depending on the outcomes they generate.

From the database, several elementary sets can be seen. In other words, it is possible to group companies with similar profiles in terms of marketing expenses (same conditional attributes) according to four different groups.

	Conditional attributes					Decision attribute
Firms	MKT	ADV	DIS	R&D	MIS	PS
F1	Low	Low	Low	Low	Low	Low
F2	Medium	Medium	Low	Low	Low	High
F3	Low	Low	Low	Low	Low	Low
F4	Low	Low	Low	Low	Low	Low
F5	Low	Low	Low	Low	Low	Low
F6	Low	Low	Low	Low	Low	Low
F7	Medium	Medium	Low	Low	Low	Medium
F8	Low	Low	Low	Low	Low	Low
F9	High	High	Low	Low	Low	Low
F10	Low	Low	High	High	High	High
F11	Low	Low	Low	Low	Low	Low
F12	Low	Low	Low	Low	Low	Low
F13	High	High	Low	Low	Low	Low
F14	Low	Low	Low	Low	Low	Low
F15	Medium	Medium	Low	Low	Low	High
F16	Low	Low	Low	Low	Low	Low
F17	Low	Low	Low	Low	Low	Low
F18	Low	Low	Low	Low	Low	Low
F19	Medium	Medium	Low	Low	Low	Medium
F20	Low	Low	Low	Low	Low	Low
F21	Low	Low	Low	Low	Low	Low
F22	Low	Low	Low	Low	Low	Low
F23	Low	Low	Low	Low	Low	Low

Examples

Table 3.1. *Database (source: Mahapatra et al. 2010)*

As part of this study, the objective is to identify the conditions for generating the highest possible total sales value of the future product. So, we then define the subset X to be characterized as the set of companies that have achieved high sales levels for their products (decision attribute PS = high). It therefore includes the companies F2, F10 and F15, according to the database previously presented (Table 3.1).

Analogously with Figure 3.1, and reasoning by elementary sets, it is possible to define the positive region of the subset X, as well as the negative region and the boundary region.

The positive region of this subset X is therefore made up of companies with marketing expense profiles allowing the generation of high sales, *and in a particular way*. In other words, it includes the elementary sets for which the conditional attributes systematically lead to a decision where "PS = high". Thus, it would be the elementary set D, including only the company F10 (Table 3.2).

Set	Description of the elementary set					Firms that make up this elementary set
	MKT	ADV	DIS	R&D	MIS	
A	Low	Low	Low	Low	Low	F1 F3 F4 F5 F6 F8 F11 F12 F14 F16 F17 F18 F20 F21 F22 F23
B	Medium	Medium	Low	Low	Low	F2 F7 F15 F19
C	High	High	Low	Low	Low	F9 F13
D	Low	Low	High	High	High	F10

Table 3.2. *Elementary sets identified*

The negative region of the subset X is made up of companies with marketing expenditure profiles allowing them to confirm that the sales that are generated will be average or low. Therefore, it includes the subsets for which the conditional attributes systematically lead to a decision of "PS = low" or "PS = medium". These consist of the elementary sets A and C.

Finally, there remains the special case of the elementary set B (F2, F7, F15, F19). This element set has identical conditional attributes, but the decision attribute differs between companies within this set. In other words, the four companies in this set have marketing expenses that are the same, yet in some cases this leads to average sales, while in other cases, to high sales. This situation is typical of the boundary region of a subset. This elementary set therefore constitutes the frontier between the positive region and the negative region of X.

After having defined the different regions of the set X, it is possible to propose an approximation adapted to the problem to be solved (Figure 3.2).

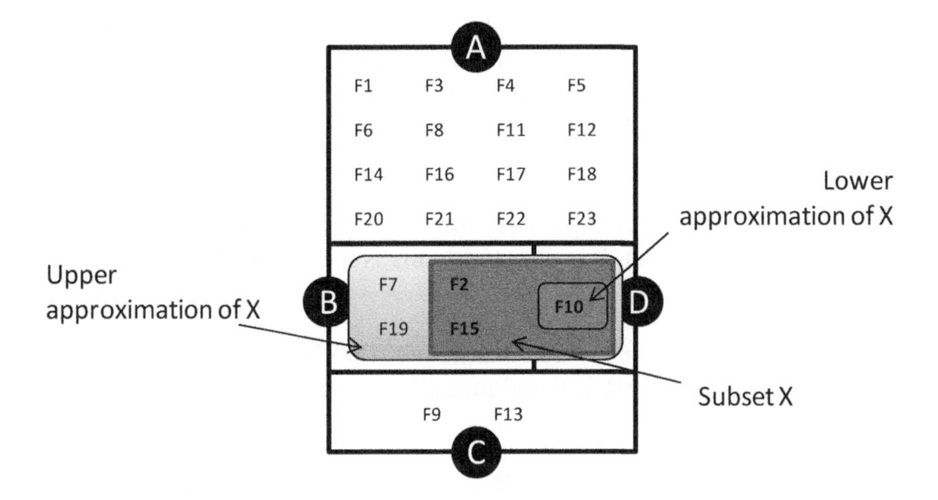

Figure 3.2. *Approximations of the subset X*

As a reminder, the lower approximation of X includes all the companies that have marketing expenses, allowing them *to ensure* a high sales volume. This is a positive region of X and therefore of the elementary set D. *According to this approximation, a company will achieve high sales if its distribution, R&D and miscellaneous expenses are high, and its marketing and advertising expenses are low.*

On the other hand, the upper approximation includes all companies with marketing expenses, allowing us to assume that they could potentially achieve high sales. This is the positive region of X together with the boundary region, and thus the elementary sets D and B. *According to this approximation, a company will potentially achieve high sales if its marketing and advertising expenses are average, and its miscellaneous, R&D and distribution expenses are low.*

3.3. Discussion of the results obtained

3.3.1. *Conditional rules obtained*

By applying the Rough Sets method to the database of Indian companies in the cosmetics sector, several rules have been identified. Each of them represents a behavioral model (combination of conditional attributes) that gives a result (decision attribute value). Based on these rules, it is possible to identify behavioral patterns that are favorable to the goal that we have set for ourselves (maximizing PS, in our case) and those that, on the contrary, are counterproductive.

First of all, only one unique decision-making rule leading to low sales was identified (PS = low). For our study, this is a behavioral model that we will try to avoid.

Rule 1: if the expenses for marketing (MKT), advertising (ADV), distribution (DIS), R&D (R&D) and miscellaneous (MIS) are low, then sales (PS) will be low:

if MKT = low, ADV = low, DIS = low, R&D = low, MIS = low,

then PS = low.

Cases supporting this rule: F1, F3, F4, F5, F6, F8, F11, F12, F14, F16, F17, F18, F20, F21, F22 and F23.

It is interesting to note that this rule is confirmed in 100% of cases within the panel of companies. Indeed, there were 16 companies that saw low sales, and each of them is defined by a spending profile that complies with this rule.

Next, two conditional rules have been connected with average sales (PS = medium).

Rule 2: if the expenses for marketing (MKT) and advertising (ADV) are average, and those for distribution (DIS), R&D (R&D) and miscellaneous (MIS) are low, then sales (PS) will be average:

if MKT = medium, ADV = medium, DIS = low, R&D = low, MIS = low,

then PS = medium.

Cases supporting this rule: F7 and F19.

Rule 3: if the expenses for marketing (MKT) and advertising (ADV) are high, and those for distribution (DIS), R&D (R&D) and miscellaneous (MIS) are low, then sales (PS) will be average:

if MKT = high, ADV = high, DIS = low, R&D = low, MIS = low,

then PS = medium.

Cases supporting this rule: F9 and F13.

In terms of the reliability of these rules, it is interesting to observe that Rule 3 is true in 100% of cases: all companies with the expense profile (MKT = high, ADV = high, DIS = low, R&D = low, MIS = low) result in average sales. On the other hand, Rule 2 is more questionable. Here, it is supported by companies F7 and F19. For these two companies, this rule is true. On the other hand, there are two other companies in the panel with the same expenditure profile (MKT = medium, ADV = medium, DIS = low, R&D = low, MIS = low): companies F2 and F15. However, the results seen by these companies were high sales of their product. Thus, rule 2 holds true in 50% of cases. Therefore, companies with an expenditure profile that follows this rule are equally likely to achieve either medium or high sales. This is explained in particular because the companies F2, F7, F15 and F19 are part of elementary set B, belonging to the boundary region of the header of X (Figure 3.2). This means the boundary region is a space where the information is incomplete, and it is necessary to approximate it. This explains the lack of reliability of this rule.

Finally, two conditional rules have been associated with high sales (PS = high). In our case, it represents a favorable behavior model, one that we would like to try to reproduce.

Rule 4: if the expenses for marketing (MKT) and advertising (ADV) are average, and those for distribution (DIS), R&D (R&D) and miscellaneous (MIS) are low, then sales (PS) will be high:

if MKT = medium, ADV = medium, DIS = low, R&D = low, MIS = low,

then PS = high.

Cases supporting this rule: F2 and F15.

Rule 5: if the expenses for marketing (MKT) and advertising (ADV) are low, and those for distribution (DIS), R&D (R&D) and miscellaneous (MIS) are high, then sales (PS) will be high:

if MKT = low, ADV = low, DIS = high, R&D = high, MIS = high

then PS = high.

Cases supporting this rule: F10.

As explained above, it should be noted that Rule 4 has the same expense profile as Rule 3, but with a different outcome in terms of sales. For its part, Rule 5 holds true in 100% of cases, but is only shown in the case of the company F10.

3.3.2. *Operational exploitation of rules*

Thus, these conditional rules give a generalized description of the knowledge derived from the panel of Indian companies in the cosmetics sector. Extracting these rules allows them to then be reused as a reference for future strategic decisions. In this way, a manager of a cosmetic company preparing to launch a new product on the market or looking to rethink its global marketing strategy will be able to comply with these rules to maximize the chances that its sales will be high. This manager will therefore seek to reproduce favorable behavioral patterns.

In view of the results, two options are available. To achieve high sales, the expense profile of the company must be the one set out in Rules 4 or 5. But as we have explained, the reliability of rules 4 and 5 varies. If Rule 5 is true in 100% of cases (although there is only one representative case in the panel), Rule 4 can lead to different outcomes in terms of sales due to the fact that it is found in the boundary region of the set (high or medium with equal probability). Thus, by selecting Rule 5, the manager chooses to take into account the lower approximation of the set, and therefore the profile that is certain to yield high sales. The choice of Rule 4 is made taking into account the upper approximation of the set, keeping in mind that it is possible (with a 50% chance) that sales will be high or that they will be average (also with a 50% chance). In general, the choice of the approximation to be taken into consideration is extremely dependent on the problem under consideration and the related context. It is a strategic decision.

Indeed, these two options reflect very differentiated strategies. Rule 4 encourages expenses on marketing and advertising, while Rule 5 favors expenses on distribution, R&D and various activities carried out by the company to enhance its image. While the first rule is more communication-oriented, the second one places more value on product performance and customer satisfaction. It is necessary for the strategy that is chosen to reflect that of the company.

On the other hand, the question of the effort required to build this spending profile should be considered. Rule 4 results in average amounts for two categories of expenses, and low amounts for three others. Rule 5 results in high expenditures on three categories and low expenditure on two others. Thus, the overall budget required and the financial effort to be made for the company are not the same. In addition, certain categories of expenses are potentially more difficult to increase and can result in direct impacts on the internal organization of the company. For example, increasing R&D expenses may involve the reorganization of this department, while increasing public expenses may result in a call for subcontracting, with no internal impact for the company.

Thus, three strategic parameters are to be taken into account in making decisions:

– Does the expense profile fit with the strategy and the image it wishes to convey?

– Is the company able to carry out this expense profile over the long-term? A large company, by its nature, can mobilize more resources than a small or mid-size business.

– Based on the previous two points, what is the best choice of approximation? Should we take the risk of having a 50% chance that sales will not be high but average if this spending profile is easier to implement operationally, or does it fit the business better?

These rules should therefore be considered as a decision support framework, translating the information hidden within a database. They are not instructions to be made; they represent behavioral models that the company will try to reproduce in order to achieve the expected results. Their formulation makes it possible to express facts in a natural and understandable language that allows them to be exploited. On the other hand, it is important to note that these rules are built from a database. To obtain precise and well-suited rules, it is necessary for this database to be representative of the context in which the decisions are made. Therefore, the rules obtained in this study can only be applied in the case of a company in the cosmetics sector, and obviously within a geographical area that is identical. The case study that we have examined does not give additional details regarding the type of companies considered, but it would also be interesting to look at the size of these companies to gain specificity in how the rules are constructed.

3.3.3. *To go further: dominance-based rough sets approach*

The dominance-based rough sets approach (DRSA) is an extension of the Rough Sets theory devoted to multi-criteria decision analysis, proposed by Greco and colleagues in 2001 (Greco et al. 2001). The main difference from the classic theory of Pawlak is the substitution of the relationship of indiscernibility with a dominance relationship, which makes it possible to deal with inconsistencies related to the consideration of criteria and the decision in order of preference.

The notion of domination refers to the definition given by Pareto (1887). In the sense of Pareto, one potential solution (decision, alternative, scenario, etc.) dominates another if it is better or equal for all the criteria taken into account, and exclusively better for at least one of them.

Thus, in the specific case of a DRSA, the classification of the examples is based on the domination relationship that binds them together. The elementary sets are arranged in order of preference, so that an example belonging to a lower subset cannot dominate an example belonging to the higher subset. If this case arises, then the example cannot be described accurately; thus, it belongs to the boundary region.

This approach brings to light two main advantages:

– first of all, it is possible to formulate more inclusive rules by using turns of phrases such as “at least” and/or “at best”;

– on the other hand, it is possible to highlight inconsistencies in the event that an example belonging to a lower subset dominates an example of the upper subset.

Now, we will look at an example to illustrate these advantages. Our application case is based on attributes evaluated according to the low/medium/high qualitative measurement scale. This measurement scale implicitly and obviously contains an order of preference: we seek to maximize this evaluation.

In the case of the decision attribute PS: it would seem obvious that the company is seeking to maximize its sales. The order of preference is therefore: low < medium < high.

In the case of conditional attributes, it is necessary to consider the choice of this order of preference. We can assume that expenses represent an important performance, although this is not obvious. Based on this premise, it seems logical to consider that high expenses are therefore better than low expenses in terms of performance. On the other hand, from a purely operational point of view, a company will seek to reduce its expenses. This means high expenses are a risk for the company. We will choose this example to consider conditional attributes from the point of view of performance. The issue of resource minimization will be brought in at the time of the final decision. The order of preference is then: low < medium < high.

This immediate order of preference of low < medium < high is intuitively taken into account when reading the rules identified, but it is not built into the initial Rough Sets method. The application of the DRSA makes it possible to refine this analysis by specifying the conditional rules and identifying certain inconsistencies.

As explained above, it appears that the companies F2, F7, F15 and F19, having the same conditional attributes – MKT = medium, ADV = medium, DIS = low, R&D = low, MIS = low – result in different decision attributes – PS = medium or PS = high. On the other hand, the dominance neutralizes this problem by allowing for the formulation of inclusive rules. Thus, it is possible to merge Rules 2 and 4 into one and the same rule, *Rule 6*: if MKT = medium, ADV = medium, DIS = low, R&D = low, MIS = low, then PS = *at least* medium.

On the other hand, it is possible to identify inconsistencies within the data table (Table 3.1). The main inconsistency that can be identified relates to the Rules 3 and 6 (ex-rules 2 and 4):

– Rule 3: if MKT = high, ADV = high, DIS = low, R&D = low, MIS = low, then PS = medium.

– Rule 6: if MKT = medium, ADV = medium, DIS = low, R&D = low, MIS = low, then PS = *at least* medium.

By comparing these two rules, it can be determined that an average financial effort on marketing and advertising leads to *medium or high* sales, whereas a high financial effort on these same attributes (and therefore a greater effort) leads to average (and therefore potentially lower) sales. Thus, if a company spends a mid-level amount on MKT and ADV, it has a high chance of obtaining high sales, but if it spends too much, these sales will decrease to reach, at best, an average level.

As established by Pareto, the company F9 following Rule 3 dominates the company F2 following Rule 6 because its expenses in MKT and ADV are higher, which translates into better performance (according to our hypothesis). However, the sales of company F9 are average, while those of company F2 are high. In view of the established order of preference, this is an inconsistency that the initial Rough Sets method does not allow to make visible. By applying the DRSA to this panel of companies, the ones that highlight this inconsistency would be excluded from the calculation and construction of the rules. This hidden information is therefore made visible by the principle of domination inherent to the DRSA. This inconsistency should be considered with greater attention. Indeed, this can be translated in different ways.

First, it may be an error in the database. In this case, it is necessary to check this value or discard it if there are any uncertainties.

On the other hand, it can be seen as a manifestation of the notion of a marketing expense threshold. After attaining a certain value, the performance no longer increases, and this no longer impacts the sales of products. It can even have a negative effect. In the particular case of marketing and advertising expenses, this can result in a "burnout" effect with customers.

Finally, it may call into question the premise that high expenses translate into superior performance. This raises the question of the efficiency of the resources used and the impact they have.

The DRSA is thus an extension of the initial Rough Sets theory, and is particularly relevant in the case of decision support problems, where the notion of preference (and therefore domination) is at the heart of the decision-making process. In fact, it is essential to be able to construct/model the preferences of a decision-maker. Very often in decision-making support, this information must be given in the form of parameters of the preference model, such as importance weightings, or in the form of various functions or thresholds. However, providing such information requires a great cognitive effort on the part of the decision maker, and it is generally accepted that it is easier to make a decision than to explain it according to specific parameters (Greco et al. 2001). This makes the DRSA particularly attractive given that it works on the basis of an exemplary decision, that is, a ranking/order of preference carried out a priori by the decision-maker. This exemplary decision therefore serves as the basis for the construction of a preference model by integrating any inconsistencies or particularities specific to the decision-maker. This preference model, which ultimately manifests itself as a set of rules, is then easily usable since it is understandable and transparent for the decision-maker.

3.4. The Rough Sets method: instructions for use

3.4.1. *The case of extracting information from a database: an example using the freeware program 4Emka*

The case study of Mahapatra et al. (2010) gives a step-by-step application of the Rough Sets method with the objective of analyzing information from a limited database (made up of 23 companies) in order to extract conditional rules in terms of marketing expenses, in order to maximize the sales of products in the cosmetics sector.

To provide a more detailed look in this section, we will now show how the existing software can assist in analyzing data. In the case where the database is limited in size and there are few conditional attributes, it is in fact quite conceivable to "manually" apply the Rough Sets method (and more particularly, the DRSA method), as the mathematical calculations are relatively basic. However, in some cases, it is necessary to process a larger database size, or a greater variety of conditional attributes. It is thus necessary to use a computer to assist in the implementation of the method. Several software programs have been developed and are available for free: RSES[1] (Rough Set Exploration System), ROSE2 (Rough Sets Data Explorer), JAMM, 4Emka, JMAF[2], etc. These programs offer different functionalities, ranging from data classification to the use of conditional algorithms for the formalization of rules.

For example, the program 4Emka, developed by the intelligent decision support systems laboratory of the Institute of Computer Sciences (Poznan University of Technology, Poland) is based on the DRSA and includes different functionalities (Gatnar et al. 2005):

– it qualitatively estimates the ability of attributes to approximate the classification of data;

– it finds the minimum attribute core (i.e. what minimum attribute reference set allows the data to be classified?);

– it sets rules for decision-making using the format "if [conditions], then [decision]", using the Domlem and Allrules algorithms;

– it applies the decision-making rules to reclassify new data.

This software has been used particularly within the context of an innovation issue related to the management of collaborative projects carried out between technology transfer centers and industrial entities (Kooli-Chaabane 2010). The DRSA methodology was used to highlight the best practices to be implemented in order to promote the success of a technology transfer. In particular, it demonstrated the importance of design boundary objects (BO) as key indicators influencing the dynamics of an innovation project and on its ultimate success. These objectives can be defined by intermediate results materialized and shared within the project team: prototypes, models, specifications, standards, technical sheets, etc.

To illustrate the use of this software, we will now return to our application case from section 3.2. We will use a simplified version of this problem by using the same conditional attributes (MKT, ADV, DIS, R&D, MIS) and the same decision attribute (PS), but we will reduce the sample

1. Available at: www.roughsets.org/roughsets/software.

2. Available at: www.idss.cs.put.poznan.pl/site/4emka.html.

size of companies by considering only the top 10 from the database given earlier.

Firms	MKT	ADV	DIS	R&D	MIS	PS
F1	Low	Low	Low	Low	Low	Low
F2	Medium	Medium	Low	Low	Low	High
F3	Low	Low	Low	Low	Low	Low
F4	Low	Low	Low	Low	Low	Low
F5	Low	Low	Low	Low	Low	Low
F6	Low	Low	Low	Low	Low	Low
F7	Medium	Medium	Low	Low	Low	Medium
F8	Low	Low	Low	Low	Low	Low
F9	High	High	Low	Low	Low	Low
F10	Low	Low	High	High	High	High

Table 3.3. *Reduced database*

To use 4Emka, the data involved in the decision-making process must be imported in the form of a text file (.txt) or data file (.isf). This file must follow the template shown in Figure 3.3.

The "Attributes" section allows you to define the criteria considered and specify their evaluation scale. For functional questions related to the use of 4Emka, we have transformed the earlier scale using words (low, medium, high) into a numerical scale, where the value 1 indicates a low evaluation, the value 2 a medium one and the value 3 a high one. We will therefore use the notation "[1,2,3]" to translate this scale, but it is also possible in the case of a purely numerical evaluation to define the scale as continuous. In this case, the notation "(continuous)" is used. Next, it is necessary to identify the decision attribute to be considered (in our case, PS).

The "Preferences" section allows you to specify the objective associated with each attribute. In our case, all the attributes are to be maximized, so we choose the "gain" option. But it is also possible to choose the options "cost" (in the case of minimization) or "none" if no goal is associated with the attribute.

```
**ATTRIBUTES

+ MKT:  [1,2,3]
+ ADV:  [1,2,3]
+ DIS:  [1,2,3]
+ R&D:  [1,2,3]
+ DIV:  [1,2,3]
+ PS    [1,2,3]

decision: PS

**PREFERENCES
MKT: gain
ADV: gain
DIS: gain
R&D: gain
DIV: gain
PS  gain

**EXAMPLES

1    1    1    1    1    1
2    2    1    1    1    3
1    1    1    1    1    1
1    1    1    1    1    1
1    1    1    1    1    1
1    1    1    1    1    1
2    2    1    1    1    1
1    1    1    1    1    1
3    3    1    1    1    1
1    1    3    3    3    3

**END
```

Figure 3.3. *Import file*

Finally, the "Examples" section brings together all the companies that we consider in our sample, as well as their evaluations using the five conditional attributes and the decisive attribute.

Once this file has been imported into 4Emka, it is possible to view the summary of attributes or examples using the "Show" tab (Figure 3.4).

Then, with the "Calculate" tab, the set of rules can be calculated and regrouped in the form of a summary table (Figure 3.5). The behavioral patterns leading to a high PS ("PS of at least 3") are then shown: Rules 7, 8 and 9. We are also shown the behavioral models resulting in a PS that is, at best, low ("PS at most 1"): Rules 1–6.

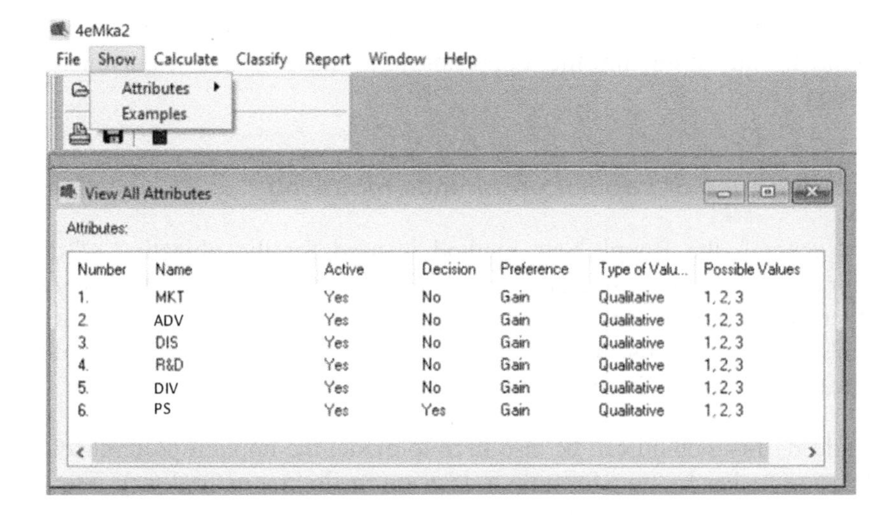

Figure 3.4. *View of attributes*

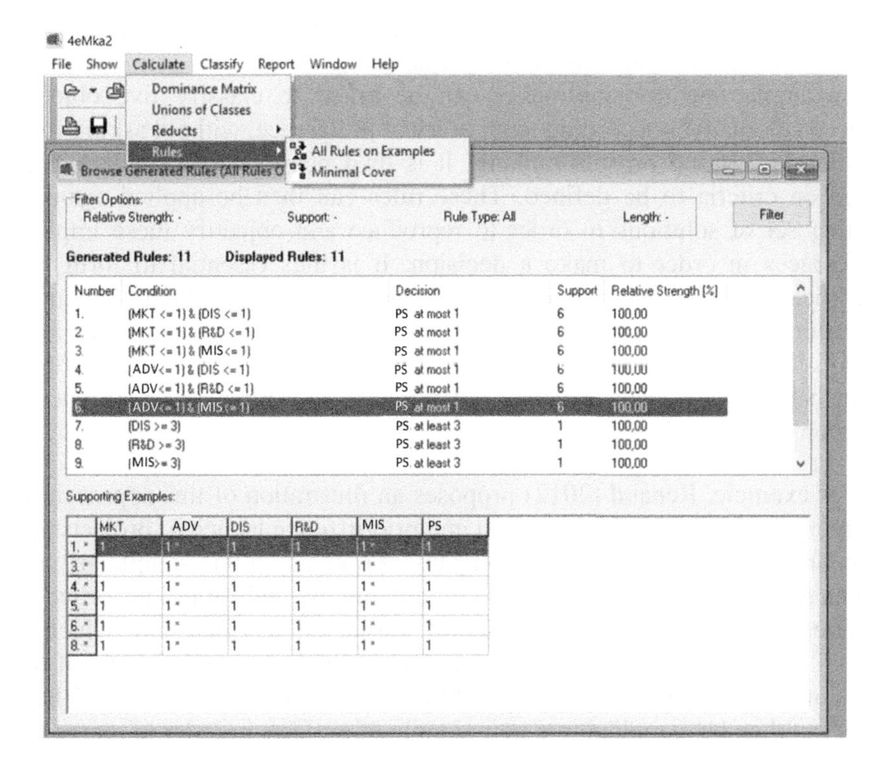

Figure 3.5. *View of rules obtained*

The software also enables visualizing the examples supporting the rules presented, thus allowing the opportunity to identify the companies in question.

3.4.2. *The case of implicit preference modeling*

In general, the Rough Sets method is used for the objective of data exploration, as in our case study, potentially (but not necessarily) relying on mathematical calculation software. We then try to identify the behavioral models that allow us to achieve a desired result. However, another application of Rough Sets (or more particularly of DRSA) is conceivable.

Indeed, this method can be also used to model the implicit preferences of a decision-maker by transforming a decision made in a natural way into the form of operable and generalizable rules. It is actually easier to ask a decision-maker to make a decision instead of explaining it. Thus, starting from an exemplary decision made a priori by the decision-maker, it is possible to determine the implicit decision-making mechanism that the decision-maker has implemented by translating it into the form of rules. For example, the decision-maker can be asked to classify five objects/projects/alternatives according to an order of preference, without asking for a justification. Based on this ranking, it is then possible to formulate rules based on criteria to be defined. These rules can then be applied again to another set of solutions in order to reproduce and objectify these implicit preferences in order to make a decision. It is thus essential to form this classification because it constitutes the input data on which the entire procedure will be based. Therefore, we are not trying to identify behavioral models that make it possible to achieve a desired result, but we are trying to build behavioral models that make it possible to explain and reproduce a decision made implicitly.

For example, Renaud (2012) proposes an illustration of this approach by applying the DRSA method to select and prioritize the launch of projects in a pulp and paper manufacturing company. From an initial sample of five products, the decision-maker gives a ranking by indicating its order of preference, without further explanation. Based on this decision, which is called "exemplary", it is possible to construct the behavioral models of the decision-maker in the form of rules of preference and non-preference. These rules can then be reapplied to a new sample of projects in order to reproduce the implicit decision of the decision maker, which has been modeled in the form of rules.

We recommend moving toward an application of this type in the event that the following conditions are met, in order to facilitate the course of the process:

– the exemplary decision (phase 1) is built from a broader view and results in the ranking a priori from a limited number of pre-selected examples. For example, the decision-maker can be asked to classify five objects/projects/alternatives according to an order of preference, without asking for a justification. This a priori ranking is primordial because it reflects the implicit preferences of the decision-maker, which we will then try to explain;

– the number of conditional attributes describing the examples is small (with a maximum of 4 or 5), so as not to exceed a reasonable cognitive load;

– the number of examples constituting the database is "reasonable" (between 5 and 20).

In the event that these conditions are not met, it is better to make use of software to apply this method.

3.4.3. *Step-by-step application*

Traditionally, the approach to modeling a decision-maker's preference follows six main steps:

(1) exemplary decision of the decision-maker: on the basis of either the definition of a decision attribute ordered as passable < good < very good, or from a ranking of a sample of examples made beforehand and spontaneously by the decision-maker;

(2) definition of conditional attributes and creation of the database;

(3) pairwise comparison of the examples constituting the database: *identification of dominated/dominant examples*;

(4) identification of rules: *formulated in the form "If... then..."*;

(5) confirmation of the rules;

(6) application of the rules to an extended data sample.

3.4.3.1. *"A priori" ranking*

Starting on the basis of a preliminary ranking carried out spontaneously by the decision-maker, the objective is to model the decision-maker's implicit preferences (in the form of a ranking) and then explain and formalize them based on rules of preference and non-preference. These rules can then be applied again to another set of solutions in order to reproduce and objectify these implicit preferences for the purpose of making a decision. It is thus essential to form this classification, because it constitutes the input data on which the entire procedure will be based.

Let us look at an example to show how this process works. Company X carried out four projects in year N. These projects were informally prioritized by the company's steering committee. That is to say, financial, human and material resources were implicitly distributed unevenly among these projects in order to favor those that were considered a higher priority and more strategic. The head of the company was then asked to rank the four projects from year N according his/her order of preference (in terms of launch priority).

This ranking was the following: project 2 > project 1 > project 3 > project 4.

Based on this preliminary ranking, the DRSA method makes it possible to characterize and work out the implicit preferences of the company in terms of a strategy for the selection of projects in the form of rules of preference and non-preference that can be reapplied for the selection of the following year's projects. We then build behavioral models that help to better understand how decisions are made and make them easier to replicate.

3.4.3.2. *Definition of conditional attributes*

The next step is to build the database to be used to formalize the decision-maker's preferences prompted by the ranking made previously. It is therefore necessary to identify the attributes that are as relevant and exhaustive as possible to characterize the examples constituting the database. These attributes or evaluation criteria must be precisely defined in order to guarantee the objectivity of the approach.

So, within the framework of our example, the four considered projects were evaluated by experts according to three criteria: the cost of implementation, the novelty of the concept and the associated risk:

– the cost was assessed from all expenses made within the framework of this project, from the design phase to its implementation. Thus, this criterion should be minimized;

– novelty and risk were assessed on a scale of 1–10. According to this scale, the value of 10 represents the best option, and therefore the highest novelty and the lowest risk.

The resulting database, known as the "evaluation matrix", is presented in Table 3.4.

Project	Cost	Novelty	Risk
Project 1	€50,000	6	7
Project 2	€70,000	6	8
Project 3	€70,000	3	5
Project 4	€100,000	9	5

Table 3.4. *Portfolio projects of the year N*

3.4.3.3. *Two-by-two comparison of projects: matrix of preference and non-preference*

From the evaluation matrix, it is then possible to make a two-by-two comparison of the elements of the database in order to build the existing dominance relationships. For this stage, it is critical to strictly follow the "a priori" ranking established by the decision-maker in order to re-transcribe the implicit preferences in this ranking within the two-by-two comparisons.

CAUTIONARY NOTE 1.– The preliminary ranking plays a crucial role in the application of the Rough Sets method. The series of comparisons to be carried out in order to build the rules of preference and non-preference is based on this classification, since it reflects the preferences of the decision maker that are to be explained. Thus, the order of the comparisons will strictly follow this ranking. In this way, the example with the highest ranking will be ranked before the second ranking, then the third, and so on. Then, the second in the ranking will be compared with the third, then with the fourth, etc.

To perform these pairwise comparisons, the two examples are compared on each of the conditional attributes. If the first example dominates the second, the value 1 is assigned. Otherwise, it is assigned the value of 0. In the event that the comparison results in both being equal, the value 0 is also assigned. Thus, a matrix is obtained consisting of 0 and 1 and called a preference matrix (Figure 3.6).

CAUTIONARY NOTE 2.– Look out for the comparisons made in the case of a criterion to be minimized. It is necessary to take into account the dominance relationships in the context of minimizing the value, and not by focusing on mathematical values. For example, in the case of the "cost" criterion, the dominant example is the cheapest project.

CAUTIONARY NOTE 3.– In the case of equal values in a comparison, the value of 0 is automatically applied.

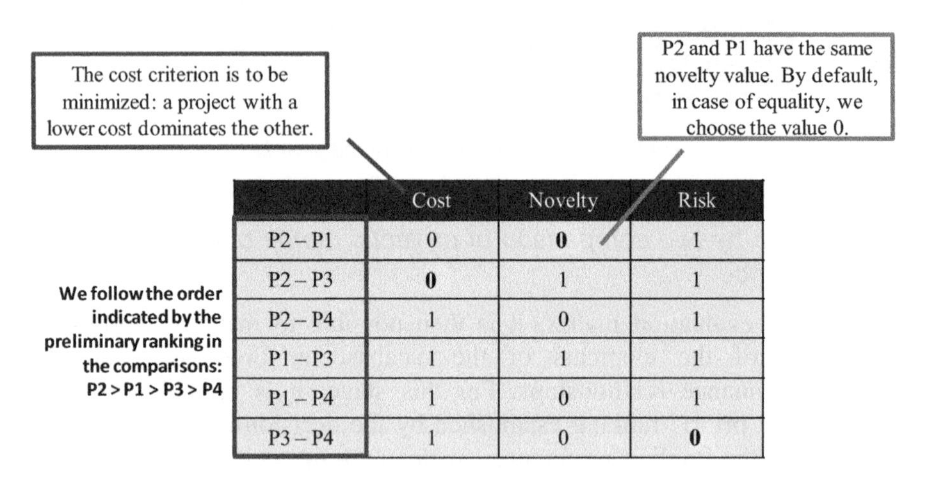

	Cost	Novelty	Risk
P2 – P1	0	**0**	1
P2 – P3	**0**	1	1
P2 – P4	1	0	1
P1 – P3	1	1	1
P1 – P4	1	0	1
P3 – P4	1	0	**0**

Figure 3.6. *Preference matrix*

The same operation is then repeated, this time to build the non-preference matrix (Figure 3.7). To construct this matrix, we reverse the order of the ranking. Thus, the comparisons are reversed.

The equality values remain unchanged, **while all other values are replaced.**

We follow the order indicated by the preliminary ranking in the comparisons: P2 > P1 > P3 > P4

	Cost	Novelty	Risk
P1 – P2	1	**0**	0
P3 – P2	**0**	0	0
P4 – P2	0	1	0
P3 – P1	0	0	0
P4 – P1	0	1	0
P4 – P3	0	1	**0**

Figure 3.7. *Non-preference matrix*

3.4.3.4. *Identification of rules*

The two-by-two comparisons of the examples have revealed binary codes distributed throughout the preference and non-preference matrices. These binary codes are the origin of the formulation of rules of preference and non-preference. This formulation requires a work of reduction and fusion of these binary codes, keeping only those that are significant.

The first step is to identify whether in one or another of the matrices, respectively, there are redundant binary codes. Binary codes appearing multiple times in the same matrix are then merged into one single rule. If they appear several times, this indicates that the rule is important. It can be determined to be a strong rule.

The second step consists of identifying whether certain binary codes appear simultaneously within both matrices. These binary codes are then discarded because they are considered inconsistent and contradictory. Indeed, an identical rule cannot reflect both the preference and the non-preference of the decision-maker.

Let us apply these two steps to the matrices obtained previously (Figure 3.8).

Preference matrix

Cost	Novelty	Risk
0	0	1
0	1	1
1	0	1
1	1	1
1	0	1
1	0	0

Non-preference matrix

Cost	Novelty	Risk
1	0	0
0	0	0
0	1	0
0	0	0
0	1	0
0	1	0

Figure 3.8. *Identification of rules. For a color version of this figure, see www.iste.co.uk/enjolras/decisionmaking.zip*

3.4.3.4.1. Step 1: redundancy

Within the preference matrix, one binary code is redundant (101). This code can therefore be merged into a single rule.

Within the non-preference matrix, two binary codes are redundant (000) and (010). They can also be merged.

These three binary codes therefore translate rules which seem to be more important than the others because they appear several times within the comparisons.

3.4.3.4.2. Step 2: inconsistency

If we observe the two matrices simultaneously, a binary code appears both in the preference matrix preference and in the non-preference matrix. This binary code (100) is therefore considered inconsistent, since it translates both a preference and a non-preference rule at the same time. Thus, it must be removed.

Finally, the rules created for this example are presented in Table 3.5.

They can be translated as follows:

– preference:

- (001): if a project is not as good in terms of cost and novelty, but is better in terms of risk, then it is preferred;

- (011): if a project is not as good in terms of cost and better in terms of novelty and risk, then it will be preferred;

- (101): if a project is better in terms of cost and risk, and not as good in terms of novelty, then it will be preferred (note that this rule appears to be strong as it is redundant);

- (111): if a project is better for all three criteria, then it will be preferred;

– non-preference:

- (000): if a project is not as good in terms of all three criteria, then it will not be preferred;

- (010): if a project is not as good in terms of cost and risk, and better in terms of novelty, then it will not be preferred (note that this rule appears to be strong as it is redundant).

Preference rules	Non-preference rules
001	000
011	010
101	
111	

Table 3.5. *Summary of the rules identified*

With these rules in mind, the company should seek to limit the risk associated with a project first (001 P), and then limit its cost (101 P). It would be able to favor a more expensive project if it is less risky, (011 P) but the novelty of a project alone does not appear to be a high-priority determining criterion (010 NP). The company therefore appears to support its strategy of prioritization over prudence.

From a theoretical point of view, the rules of preference and non-preference provide a model for the implicit decisions of the decision maker. If we consider the subset X of the projects preferred by the decision-maker, then the preference rules derive from the positive region of this set, while the negative region is reflected in the non-preference rules (Figure 3.9). The boundary region of set X is represented by the rules that have been discarded, namely the inconsistent rules present both in the preference matrix and the non-preference matrix.

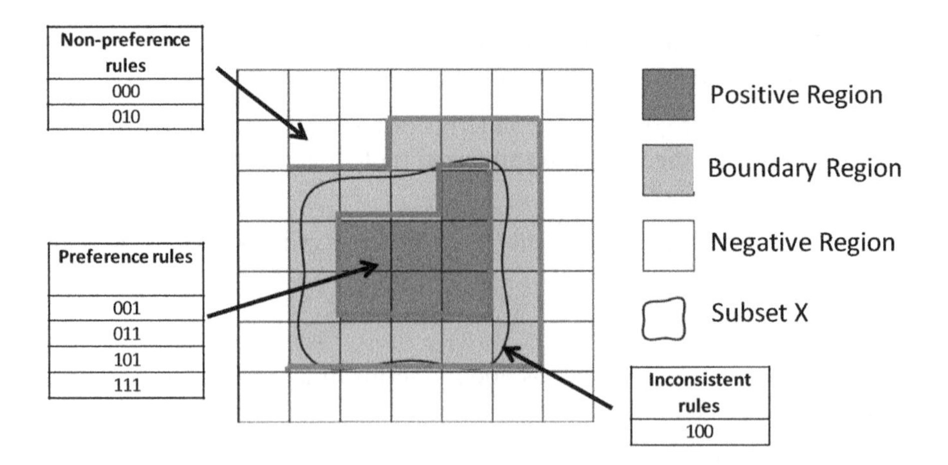

Figure 3.9. *Illustration of positive and negative regions*

3.4.3.5. *Validation of the rules*

The final stage of creating the rules consists of verifying that the binary codes chosen correctly reflect the initial preferences of the decision-maker. To do this, it is necessary to validate the rules by reapplying them to the initial sample. Each comparison of the examples will then be carried out again and associated with a corresponding rule. In the event that the rules have been defined correctly, they can be used to recover the initial ranking of the decision-maker:

1) Within a table, make all possible comparisons and identify the related binary code.

2) In the case where the related binary code corresponds to a known preference rule, assign the value +1 to the first example of the comparison and -1 to the second example.

3) In the case where the related binary code is for a known non-preference rule, assign the value -1 to the first example of the comparison and +1 to the second example.

4) In the case where the binary code obtained is for a rule that has been removed because it is inconsistent, do not assign any value.

5) Add together all the values from each example. The example with the highest sum is the first in the ranking, and so on.

In Figure 3.10, it can be seen that the initial ranking is returned: P2 > P1 > P3 > P4. The rules are therefore valid because they make it possible to classify the initial sample following the order indicated by the decision-maker.

		Associated binary code	Type of rule	P1	P2	P3	P4
Preference matrix	P2 – P1	001	P	-1	+1		
	P2 – P3	011	P		+1	-1	
	P2 – P4	101	P		+1		-1
	P1 – P3	111	P	+1		-1	
	P1 – P4	101	P	+1			-1
	P3 – P4	**100**	/				
Non-preference matrix	P1 – P2	**100**	/				
	P3 – P2	000	NP		+1	-1	
	P4 – P2	010	NP		+1		-1
	P3 – P1	000	NP	+1		-1	
	P4 – P1	010	NP	+1			-1
	P4 – P3	010	NP			+1	-1
	Sum			+3	+5	-3	-5

Figure 3.10. *Validation of the rules*

CAUTIONARY NOTE 4.– In some cases, the step of checking the rules does not allow to scrupulously determine the initial ranking. Equal values or slight changes in the ranking may be found. This is due, among other reasons, to the potential loss of information when excluding the conflicting rules identified. The initial sample used to construct the rules may therefore not be sufficiently significant to reflect the preferences of the decision-maker. One option may be to expand the initial sample to make it more representative.

3.4.3.6. *Application of the rules on a new sample*

The objective of the DRSA is to identify rules that translate the implicit preferences of a decision maker. Thus, the last step to be carried out is to implement these rules, applying them to a new sample in order to obtain a ranking for these new examples that faithfully reflects the preferences of the decision maker. The goal is therefore to reproduce the behavioral models identified for this type of decision.

In the case of our example, a possible implementation of the rules could be to reapply them to the project portfolio for the year N+1 in order to identify the projects to be prioritized for the coming year.

Company X is planning three projects for the year N+1. They were evaluated by an expert committee, as shown in Table 3.6.

Year N+1	Cost	Novelty	Risk
Project A	€150,000	7	7
Project B	€90,000	4	6
Project C	€40,000	9	5

Table 3.6. *Portfolio of projects for the year N+1*

In order to create the ranking of the projects for the year N+1, it is necessary to make all possible comparisons between projects A, B, and C, and to generate the corresponding binary codes. These binary codes will potentially create rules which have been identified earlier, and thus represent known behavior patterns:

(1) in the case where the related binary code represents a known preference rule, assign the value +1 to the first example of the comparison and −1 to the second example;

(2) in the case where the related binary code is for a known non-preference rule, assign the value −1 to the first example of the comparison and +1 to the second example;

(3) in the case where the binary code obtained is for a rule that has been removed because it is inconsistent, do not assign any value.

Finally, add together all the values from each example. The example with the highest sum is the first in the ranking, and so on (Table 3.7).

	Associated binary code	Type of rule	A	B	C
A – B	011	P	+1	-1	
B – A	100	/			
A – C	001	P	+1		-1
C – A	110	/			
B – C	001	P		+1	-1
C – B	110	/			
Sum			+2	0	-2

Table 3.7. *Application of the rules to the projects for the year N+1*

CAUTIONARY NOTE 5.– When applying the rules that have been generated to a new sample, new binary codes appear during the comparisons. These cannot be interpreted because they do not correspond to rules that are known and identified as a behavioral model. In this case, the row associated with this comparison is left blank. For example, for comparisons C–A and C–B, the associated code 110 is unknown. It represents neither a previously identified rule of preference, nor a rule of non-preference, and does not translate a contradictory rule either.

Applying these steps as part of our project prioritization example, it appears that for the year N+1, project A is preferred first, then project B and finally project C. It is interesting to note that this ranking follows the order of the lowest risk. Project A is in fact the least risky, yet it is also far and away the most expensive. Project C, though very inexpensive compared to the others and much more innovative, is ranked poorly, despite the fact that its risk assessment is only two points below that of project A. This is explained in particular because the re-application of the rules obtained only translates the rule of preference (001), which favors risk (see the binary code column from Figure 3.8). The combinations carried out do not reveal any other known rules (110 – unknown rule, and 100 – contradictory rule). In this configuration, there is little room for nuance, which explains this very stark differentiation in the ranking. However, the notion of nuance appears to be critical, since in this example, there is some room for consideration about the importance of the difference between two projects. Projects A, B and C are relatively close in terms of risk (7, 6, and 5, respectively, on a scale of 1 to 10). Will a difference of one point between two projects really be significant for the decision maker? How big of a difference is needed for the preference to clearly be higher? The same comment could be made about the cost. Proportionally, the cost gaps between projects are much higher. However, this is not taken into account by the DRSA. A difference of one point and a difference of 10 points are both considered in exactly the same way. This issue is therefore important to take into account in some cases. One possible option would be to propose preference functions for each criterion to add nuance to the decisions (see Chapter 1).

ROUGH SETS AT A GLANCE.– Objective: the Rough Sets method is a descriptive method. Depending on the context of its application, it allows (1) to extract behavioral models within a database, allowing an intended goal to be reached and (2) to build behavioral models allowing a decision made intuitively to be explained and to reproduce it. One of its peculiar features is as follows: this method does not require the use of weighting for the criteria, as is the case in most of the other methods shown in this book. It does not create a ranking of the alternatives; on the contrary, the ranking is given in the input data that we are trying to explain.

Limitations: this is a method that is difficult to understand conceptually. It is similar to the "fuzzy logic" that we will examine in Chapters 5 and 6 of this book.

3.5. References

Adams, P., Bodas Freitas, I.M., Fontana, R. (2019). Strategic orientation, innovation performance and the moderating influence of marketing management. *J. Bus. Res.*, 97, 129–140.

Booms, B.H. and Bitner, M.J. (1980). New management tools for the successful tourism manager. *Ann. Tour. Res.*, 7, 337–352.

Gatnar, E., Rozmus, D., Bock, H.-H., Gaul, W., Vichi, M. (2005). Data mining – The Polish experience. In *Innovations in Classification, Data Science, and Information Systems. Proceedings of the 27th Annual Conference of the Gesellschaft Für Klassifikation e.V., Brandenburg University of Technology, Cottbus, 12–14 Mars, Studies in Classification, Data Analysis, and Knowledge Organization*, Baier, D. and Wernecke, K.-D. (eds). Springer, Berlin Heidelberg.

Greco, S., Matarazzo, B., Slowinski, R. (2001). Rough sets theory for multicriteria decision analysis. *Eur. J. Oper. Res.*, 129, 1–47.

Kooli-Chaabane, H. (2010). Le transfert de technologie vu comme une dynamique des compétences technologiques : application à des projets d'innovation basés sur des substitutions technologiques par le brasage métallique. PhD thesis, Institut National Polytechnique de Lorraine, Nancy.

Kotler, P., Keller, K., Manceau, D. (2012). *Marketing Management*. Pearson, London.

Mahapatra, S., Sreekumar, Mahapatra, S.S. (2010). Attribute selection in marketing: A rough set approach. *IIMB Manag. Rev.*, 22, 16–24.

Nagard-Assayag, E.L., Manceau, D., Morin-Delerm, S. (2015). Les dilemmes du marketing de l'innovation. In *Le marketing de l'innovation. Concevoir et lancer de nouveaux produits et services*, Nagard-Assayag, E.L., Manceau, D., Morin-Delerm, S. (eds). Dunod, Paris.

Pawlak, Z. (1998). Rough set theory and its applications. *J. Telecommun. Inf. Technol.*, 29, 7–10.

Pointet, J.-M. (2011). Rôle du marketing en conception innovante : les leçons du cas Axane. *Gestion 2000*, 28, 65–80.

Renaud (2012). Méthode pour trouver les meilleurs produits par les cartes rough sets. Report, Renaud.

Sawicki, P. and Żak, J. (2014). The application of dominance-based rough sets theory for the evaluation of transportation systems. *Procedia – Social and Behavioral Sciences, Transportation: Can We Do More with Less Resources? 16th Meeting of the Euro Working Group on Transportation*, 111, 1238–1248.

4

Building a Coherent Project Portfolio: An Application of MAUT

Project management is an essential element in the development of a company's capacity to innovate. Increasing the pace of output of innovative projects can only be done by mastering the resources involved: financial and technical resources, but also time.

However, innovative companies are required to develop several projects simultaneously, which form a project portfolio. Since all companies have limited resources, the construction process of this project portfolio is a recurring problem in innovation. This means that it is necessary to make a selection related to the objectives of decision makers or the concerned organization, but also regarding the overall compatibility of this set of projects, to form a coherent whole. Indeed, as Belton and Stewart (2002) have found, the creation of a project portfolio requires not only giving consideration to the individual characteristics of the alternatives but also the way they interact with each other, and the positive or negative synergies that may arise.

To illustrate this problem, we will draw inspiration from the works of Lopes and Almeida[1] (2013, 2015), aimed at evaluating synergies which

1 For educational reasons, some of the results of this study have been adapted.

allow for the selection of a portfolio of projects, within the specific context of the petroleum industry. This article proposes an application of the multi-attribute utility theory (MAUT) method, relying on the theory of utility developed by Von Neumann and Morgenstern in 1953.

4.1. Context and challenges in decision-making

4.1.1. *The selection of innovative projects*

The notion of uncertainty is intrinsically linked to innovation projects; this may involve uncertainty about the technical feasibility, doubts about how it will be received on the market, whether new regulatory constraints will be put in place, etc. Thus, in innovation, simply managing the project is not enough: it needs to be led. Flexibility, reactivity and the use of tools and methods allow for the uncertainty to be limited. In addition, innovative companies have to develop several projects simultaneously, sometimes combining short, medium and long-term projects. This is known as multi-project management. Since all companies have limited resources, sometimes it will be necessary to make trade-offs.

In response to this problem, a specific management practice has been developed: project portfolio management. For this, the project portfolio is defined as a regrouping of the company's projects in such a way as to allow a balanced allocation of resources according to the needs and priorities of the company. This also allows for the centralization of project information so as to ensure consistency and adaptability to the corporate strategy. This means the selection and prioritization of the innovation projects that make up the portfolio are not based solely on a financial analysis of the return on investment, but it must also take into account strategic issues, the characteristics of the projects, the implications of R&D, the conditions of commercialization, the use of resources, the probabilities of success of the projects and the non-financial benefits arising from them.

4.1.2. *Decision-making in the petroleum industry*

The case that we have chosen to deal with in this chapter consists of studying the process of project selection in the oil industry, in order to create a coherent project portfolio. The oil industry is one of the most powerful industries in the modern economy, bringing together many different stakeholders: countries spread out over the five continents (of many different sizes and economic power levels), multi-national firms, political and environmental organizations, etc. In particular, oil exploration and production (E&P) activities are generally risky activities, which can also generate a high financial return. This industry is characterized by high volatility in the prices of its products and by high pressures to reduce cost structures as much as possible. Uncertainties can also be reflected in the volume of available hydrocarbon reserves, the influence of the environment, geopolitics, or tax structures. When it comes to oil exploration projects, it is essential to have a comprehensive analysis carried out covering the various factors and nuanced issues that come into play during the selection phase of these projects. This requires a good understanding of the preferences of the decision makers and the business environment in which they operate.

However, traditionally, decision-making in the oil industry is mainly based on financial indicators (Bailey et al. 2000; Armstrong et al. 2004). Thus, projects are evaluated according to their level of profitability, the expected production or the cost of the initial investment. Given the high level of risks and uncertainties involved in this sector, this project selection phase would benefit from a consideration of, in addition to the economic aspects, the technical, human and commercial aspects (particularly), in order to be able to assess the effective value of a project for the organization that will carry it out. Thus, the selection of a project portfolio becomes a multi-dimensional issue where all of these aspects influence the preference structure of decision makers. The selection of projects is therefore based on the notion of compromise, where the best option is the one that maximizes the objectives of decision makers, while knowing beforehand that an improvement of one aspect could potentially lead to a loss of quality for another.

In order to take into account these specific considerations, this case study is based on a multi-objective decision model created on the basis of the MAUT methodology.

4.1.3. *Definition of the decision model*

In their research paper, Lopes and de Almeida examine the case of the company Petrobras, a publicly traded company whose majority shareholder is the Brazilian government. This company is active in 28 countries and conducts activities in the exploration, production, sale and transportation of oil and natural gas. The authors focus on the exploration activities of Petrobras in order to propose a decision support method for the creation of a coherent project portfolio that creates as much value as possible. This chapter is inspired by the works of these authors. The case addressed here is based on real data, but the illustrative example given has been adapted.

The decision model was built in collaboration with the head of the company's project portfolio, with the help of a decision-making expert. They arrived at a proposal made up of three main factors, measured using quantifiable indicators:

– *The financial return on investment*, measured by the net present value (NPV) of the project. This is a measure of the profitability of an investment based on the sum of the resulting cash flows. This commonly used financial indicator can be qualified as a "natural" attribute to measure this goal of profitability; in other words, it is based on an objective measurement. However, many economic uncertainties can be raised in this context, especially in terms of costs and technology. With these considerations in mind, the calculation of the NPV of a project was carried out from numerical simulations based on economic models, in order to propose a probabilistic NPV value.

– *The expected hydrocarbon production*, measured through available reserves. In this case, the factor of available reserves (RES) is an indirect measure of the company's hydrocarbon production. Therefore, the measurement is not taken of the objective (production) directly, but instead of a factor associated with it (available hydrocarbon reserves). This indirect factor is reflected in the size of the reserves associated with each project. Again, this measure is probabilistic because it is not verifiable a priori.

– *The influence of external factors*, measured by a qualitative value reflecting the political situation of the target country and infrastructure elements (EXT). This factor is related to the fact that oil investments are made in countries with very different forms of governance and administration, as well as varied infrastructures. Thus, it is necessary to assess the risk involved. Together, these elements are then aggregated in the form of a Likert scale from 1 to 5, where 1 reflects a very negative influence of external factors on the project, while 5 reflects a very positive influence.

It is interesting to note that the proposed decision model takes into account both quantitative (NPV, RES) and qualitative (EXT) factors. In addition, the particular context of oil exploration projects necessarily implies a high degree of uncertainty for the measurement of these factors. Most of these are probabilistic measures. The MAUT method is recognized for its ability to handle stochastic data (Zionts 1992), making it an appropriate choice.

4.2. The MAUT method

The MAUT method, known as a synthesis method, was created by Ralph Keeney and Howard Raiffa in the 1960s in the United States. Like other methods of multi-criteria analysis, it is structured to address the problems of the compromises to be made between the multiple objectives of a decision. It is based on the assumption that decision makers, as part of their decision-making process, will systematically try to optimize (consciously or unconsciously) a mathematical function that aggregates the entirety of their point of view. In other words, there is a function (already known or to be determined) that represents the desirability (or preference) of an alternative, taking into account that the decision to be made can involve simultaneously many criteria. This mathematical function is called a utility function.

4.2.1. *Terminology and methodological concepts*

The terminology used in this method is based on the notions of alternatives and criteria.

The alternatives are the different elements that we are looking for to qualify and to rank. In the case that we have chosen to address (the creation of a project portfolio), the alternatives are the different projects that we are trying to select.

The criteria, on the other hand, are the factors that allow us to qualify the alternatives to which an objective has been assigned. That is to say, a criterion is a factor to which a target value has been assigned (maximization, minimization or a precise value to be achieved). These criteria are evaluated using measurement indicators, for example, in thousands of euros, on a scale from 1 to 10, in units of weight, etc.

Finally, the MAUT method is based on a fundamental notion: utility. It can be defined as the desirability (or preference) of an alternative, taking into account the multiple objective nature of the decision to be made. Therefore, there is said to be a total utility (that of the alternative) which aggregates partial utilities (the utilities associated with each criterion of the decision model). Thus, according to the founding hypothesis of MAUT, decision makers will move toward the alternatives with the highest total utility, itself characterized by a combination of the partial utilities of the criteria composing it. Utility therefore translates into a mathematical function: the utility function.

We will now look at the case of a partial utility function (meaning it relates to a particular criterion). The role of this function is, for a given alternative, to transform the value of the evaluation of a criterion into a utility value between 0 and 1 (where 0 represents no utility at all and 1 represents maximum utility). Thus, we can arrive at a function U(x) which, if applied to the set X of the values taken by the alternatives for criterion 1, transforms these values into utility values. Thus, the values {a,b,c,d} composing the set X are transformed into a set of values {U(a), U(b), U(c), U(d)}, between 0 and 1. According to Figure 4.1, the value c translates the lowest utility U(c) (close to zero), while the value d translates the highest utility U(d) (close to 1). It is also interesting to note that the values a and b reflect equal utility. From the point of view of decision makers, they are therefore equivalent and not differentiable for the decisions they make.

Thus, the utility function mathematically models decision makers' preferences by reflecting their intrinsic requirement for a criterion. Now, we will look at the example of buying a car, and more specifically the criterion related to the power of the car. Figure 4.2 represents two different utility functions for this same criterion. It is a model of the preferences of two different decision makers. Decision maker 1 has an exponential utility function, which begins to grow very rapidly after reaching 120 horsepower. Before reaching this value, its utility curve remains very low, and translates utility values close to zero. Thus, it is possible to clearly visualize that this decision maker is very demanding regarding the power criterion in the purchase of a car. The satisfaction (utility) only increases after 120 fiscal horsepower. On the contrary, decision maker 2 has a logarithmic utility function. This decision maker's preference curve grows very rapidly from the lowest values of engine power and stagnates around a utility close to 1 (and therefore, the maximum). This decision maker therefore seems less demanding regarding this criterion; a low engine power seems to be enough. Thus, for the same engine power (105 HP), decision maker 1 will be very unsatisfied (U(1) = 0.1), while decision maker 2 will be very satisfied (U(2) = 0.8).

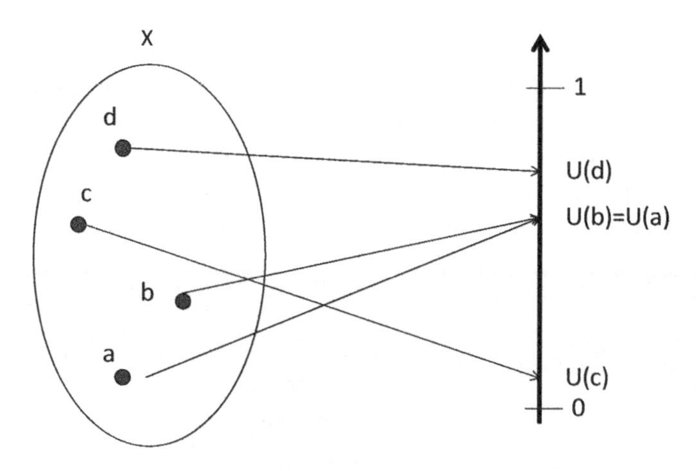

Figure 4.1. *Utility function modeling*

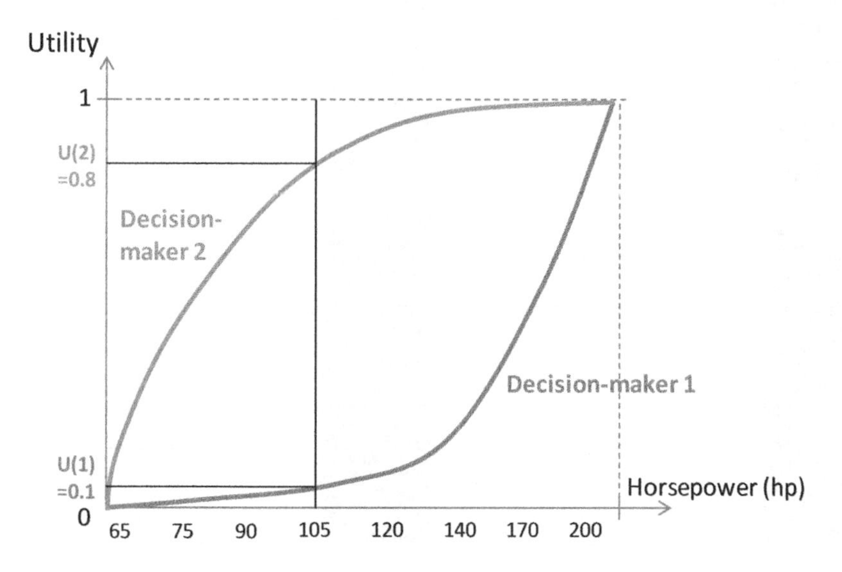

Figure 4.2. *Requirement profiles of two different decision makers*

4.2.2. *Application of the method*

In order to clarify these methodological concepts, we will apply the different steps of the MAUT method to the case study that we have chosen to examine: the selection of a portfolio of projects for the company Petrobras.

4.2.2.1. *Project evaluation matrix*

The selection of this portfolio will be made from a set of thirty proposed projects, and evaluated on the basis of the three factors that make up our decision model: NPV (in millions of dollars), RES (in millions of barrels) and EXT (on a scale from 1 to 5). These 30 projects were evaluated by experts, who then assigned them the following values presented in Table 4.1.

Projects	NPV (millions of $)	RES (millions of barrels)	EXT (1–5)	Projects	NPV (millions of $)	RES (millions of barrels)	EXT (1–5)
P1	1,086	1,311	2	P16	430	651	1
P2	670	582	5	P17	383	575	3
P3	2,131	1,710	2	P18	374	550	4
P4	991	799	3	P19	320	423	2
P5	1,172	750	4	P20	338	500	3
P6	385	512	2	P21	455	450	3
P7	1,164	850	4	P22	337	492	4
P8	1,639	1,355	1	P23	56	101	3
P9	451	678	2	P24	28	580	2
P10	829	700	1	P25	155	304	5
P11	752	708	5	P26	95	180	2
P12	457	510	5	P27	266	204	3
P13	463	480	4	P28	35	50	4
P14	709	800	1	P29	185	176	1
P15	557	850	1	P30	153	165	3

Table 4.1. *Project evaluation matrix*

4.2.2.2. *Construction of partial utility functions*

All of the proposed projects are evaluated based on the three factors: NPV, RES and EXT. However, these factors must be transformed into decision criteria. For this, it is necessary to assign them an objective, which depends on the requirements of the decision maker: is it desired for the factors identified to be systematically maximized? Would it be desirable to minimize them? Is there a target value that would be ideal to reach? The answer to these questions can be found in the utility functions associated with the criteria. To translate decision makers' preferences using each criterion, it is necessary for them to create a utility function that best models their point of view. Several methods can be used to accomplish this.

First, there is the *direct method* in which decision makers evaluate the parameters by answering direct questions about their preferences. If we consider our case study, we could first ask the decision maker to choose the project that is least suitable in terms of financial profitability (NPV) as well as the one that is the most satisfying. Once the minimums and maximums have been identified, intermediate questions can be asked. For example: "If one oil extraction project has an NPV of \$500 million and another has an NPV of \$100 million, which one do you prefer?" Or: "How would you rate your satisfaction for a project with an NPV of \$80 million?"

Based on the answers to this series of direct questions, it is then possible to identify the preference function associated with the decision maker's profile.

To do this, the simplest option is to approximate the answers of the decision maker to linear mathematical functions, as they are well known and easily identifiable. These can come in three types: linear increasing, linear decreasing and target value (Figure 4.3).

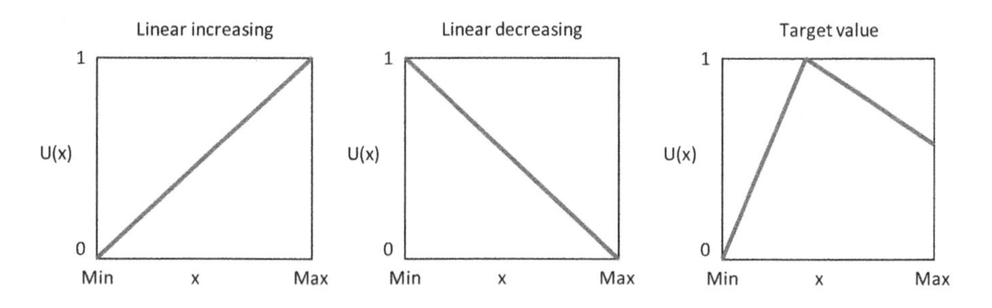

Figure 4.3. *Linear utility functions*

Another more accurate option is to conduct an ordinal regression (Figure 4.4). This involves fitting a mathematical function to a scatter plot. Here, we are looking for the utility function that best matches the data. The main advantage of this option is that it leads to more accuracy in modeling the preferences of the user. It allows us to propose more complete utility functions. However, a regression can potentially highlight several different functions with potentially close plots (e.g. an increasing linear function and an exponential one). The choice is thus very dependent on the accuracy of the answers given by the decision makers, and the amount of data available. Finally, making this choice requires the use of dedicated mathematical tools.

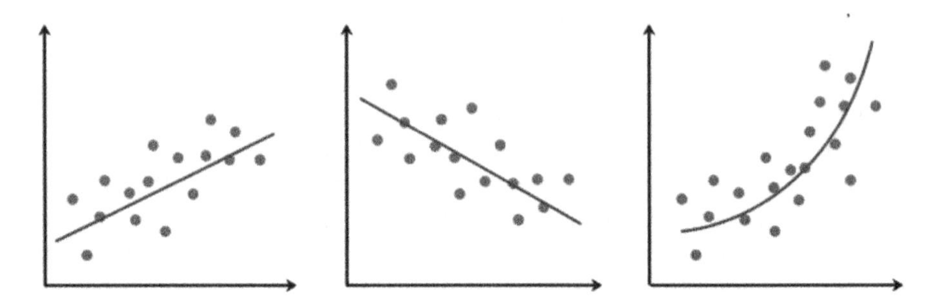

Figure 4.4. *Ordinal regressions made from a scatter plot*

On the other hand, an *indirect method* can also be used. In this method, decision makers are asked to carry out a ranking of the alternatives (by criterion), and from this ranking, the parameters of the utility function are identified. However, this method is not the most widely used because it is difficult to implement operationally.

For our case of application, a direct method of constructing the utility functions has been chosen. More specifically, increasing linear functions were assigned to the three decision criteria (Figure 4.5). This choice was made in particular for operational reasons. It was thus decided that the three decision criteria should be maximized: the higher their value, the greater their utility. The utility value 0 is the minimum value of the criterion within the set of anticipated projects, while the utility value 1 is the maximum value of the criterion.

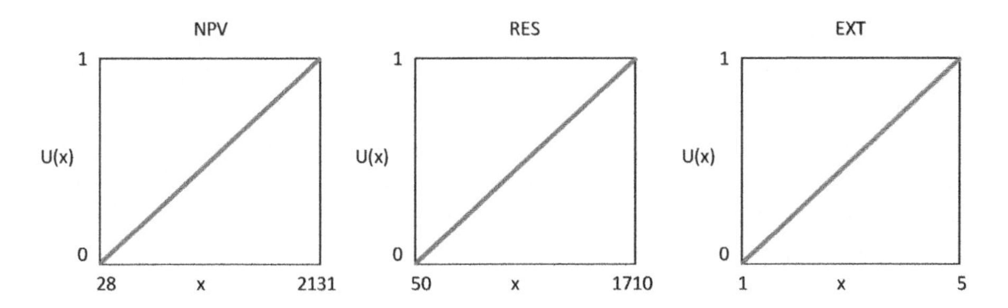

Figure 4.5. *Partial utility functions used*

In this way, the mathematical expression of these functions is as follows:

$$U_i(x) = \frac{x - min_i\,(x)}{max_i(x) - min_i(x)},$$

where:

– x is the value of the evaluation of the alternative on criterion i;

– $min_i\,(x)$ is the minimum of the evaluations of the set of alternatives on criterion i;

– $max_i\,(x)$ is the maximum of the evaluations of the set of alternatives on criterion i.

Therefore:

For NPV: $U_{NPV}(x) = \frac{x}{2{,}103} - \frac{28}{2{,}103}$

For RES: $U_{RES}(x) = \frac{x}{1{,}660} - \frac{50}{1{,}660}$

For EXT: $U_{EXT}(x) = 0.25x - 0.25$

with x being the value of the evaluation of the alternative on the criterion under consideration.

4.2.2.3. *Weighting of the criteria*

The stage of the weighting of the criteria is a common step in all aggregative decision support methods. This is a central step in the modeling of decision makers' preferences and reflects the relative importance of the selected criteria (and thus their impact on the decision). Though this step can be done declaratively, assigning a value directly ("criterion 1 is two times more important than criterion 2, etc."), intuitively distributing the "scores" ("I have 100 tokens to distribute over all my criteria"), it is preferable to rely on a constructed method, as this is more rigorous and accurate (see Chapter 2 on the AHP method, for example).

In our case, the Swing weighting procedure has been applied (Edwards and Barron 1994). This consists of asking the decision maker: "Imagine that this project has the lowest score for all the criteria taken into account. If you have the opportunity to change a single score from the worst value to the best, which criterion would you choose?" Thus, by iteration, a ranking of the criteria is obtained, ranging from the most important to the least important. However, this is only the first step, which must then be refined through the use of lotteries (probability of obtaining a profit), as advocated by Keeney and Raiffa. The combination of these two steps makes it possible to assign a weighting value to each criterion, which represents its relative importance from the point of view of the decision maker.

Inspired by this method, we can therefore determine that the weightings of the decision criteria are as follows.

NPV	RES	EXT
0.663	0.263	0.074

Table 4.2. *Weighting of criteria*

4.2.2.4. *Calculation of total utility of each project*

Once the partial utility functions as well as the weightings of the different criteria are defined, several aggregation methods can be used to construct the total utility function of each alternative.

The simplest of them is additive aggregation. The total utility U_{TOT} is then represented by a weighted sum of the utility values of each criterion. It is therefore given in the following form:

$$U_{TOT}(x) = \sum_{i=1}^{N} \omega_i U_i\,(x),$$

where:

– ω_i is the weight assigned to the criterion i;

– $U_i(x)$ is the utility value from the alternative x for the criterion i;

– N is the number of criteria.

The main advantage of this additive method is its ease of use. The utility function can thus be subdivided, but it only remains valid under certain conditions. Effectively, the criteria must be defined in such a way that they are "additively independent". Two criteria, 1 and 2, are said to be additively independent if the preference between two lotteries (defined as common probability distributions on both criteria) depends solely on their marginal probability distributions. In other words, it is considered that this independence is respected when the interactions between two criteria do not have an impact on the result of their sum. On the other hand, it is necessary to note that the additive aggregation functions necessarily imply a compensation phenomenon. Indeed, one limitation of the weighted sum is the fact that a large value of weight could compensate a smaller one. Thus, if one alternative receives strong evaluations on the majority of criteria but has a major weak point on another one, the weighted sum will compensate for this weak point. This compensation phenomenon is not necessarily a problem in decision-making, but it depends on the context. In some cases, having a weak point on a particular criterion can be a real problem. For example, in the food industry, it would be unwise to compensate a product that tastes bad with a low price or with a more enticing appearance.

Thus, when the conditions of additive independence and/or acceptable compensation are not met, other more complex methods of aggregation can be used. These include the weighted product of the criteria, where the total utility is expressed as the product of the utility values of each criterion with their weight as exponent. The Choquet integral can also be used. This method allows, among other things, to take into account the phenomena of interaction between criteria.

In the case of our study, a total additive utility function was chosen. It is given in the following form:

$$U_{TOT}(x) = \sum_i \omega_i . U_i(x)$$

where:

– $i \in \{\text{NPV}, \text{RES}, \text{EXT}\}$;

– x belongs to the set {P1, P2, ..., P30} of the proposed projects.

Projects	NPV (millions of $)	RES (millions of barrels)	EXT (1–5)	U NPV Weight = 0.663	U RES Weight = 0.263	U EXT Weight = 0.074	Total U	Ranking
P3	2,131	1,710	2	1.00	1.00	0.25	0.94	1
P8	1,639	1,355	1	0.77	0.79	0.00	0.71	2
P1	1,086	1,311	2	0.50	0.76	0.25	0.55	3
P7	1,164	850	4	0.54	0.48	0.75	0.54	4
P5	1,172	750	4	0.54	0.42	0.75	0.53	5
P4	991	799	3	0.46	0.45	0.50	0.46	6
P11	752	708	5	0.34	0.40	1.00	0.41	7
P2	670	582	5	0.31	0.32	1.00	0.36	8
P10	829	700	1	0.38	0.39	0.00	0.36	9
P14	709	800	1	0.32	0.45	0.00	0.33	10

Table 4.3. *Selection of exploration projects*

Therefore, it is a weighted sum that aggregates the three defined partial utilities of the decision criteria into a single value. This implicitly means that the criteria are considered independently of each other and that the compensation phenomenon is acceptable for this decision-making process. These assumptions will be discussed in section 4.3.

Thus, the comparison of the total utility value between one project and that of other projects makes it possible to judge its relevance in view of the preferences of the decision makers. It is then possible to carry out a ranking of the 10 projects that appear the most relevant to be included in the firm's portfolio (Table 4.3, Figure 4.6).

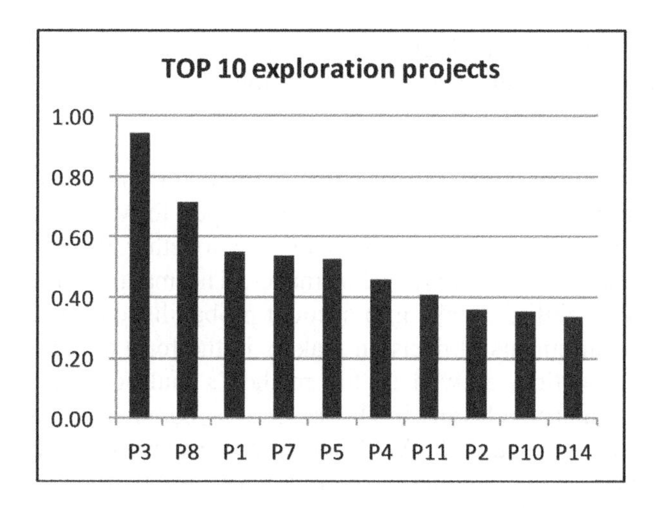

Figure 4.6. *Ranking of exploration projects*

It is interesting to note that the first two projects in this ranking (P3 and P8) contrast very clearly from the others. They are in fact the projects with the best evaluations on the NPV and RES criteria (the criteria considered to be the most important), which explains why their total utility value is very high. Then, the projects P1, P7 and P5 are very close together. All three are relatively equivalent in terms of return on investment (NPV); on the other hand, their profiles differ regarding available reserves (RES) and external factors (EXT). P1 is better positioned in terms of reserves, but will be negatively impacted by external factors, while P7 and P5 are in a positive context but with less significant reserves. Despite its unfavorable background, P1 is ranked at the top of this group, in particular because the weight assigned to external factors is low. This is also reflected in the ranking when we look at the project P8, which is ranked second with good evaluations in terms of NPV and RES, but with a big weak point in terms of external factors (a very unfavorable context). This compensation phenomenon described above is therefore particularly visible here. It is thus interesting to consider the real meaning of the weightings of the criteria in this case. According to Lopes and de Almeida, the weightings of the criteria in the case of an additive aggregation can be considered in a slightly different way than the relative importance they assume in the decision. They potentially reflect the acceptable degree of compensation that they can be given. Thus, in our case, it is considered that it would be completely acceptable to compensate for a low value on the EXT criterion due to its low weighting, while it would be unacceptable to compensate for a low NPV value because this value has been given a higher weighting. This is a way of offsetting the compensation that occurs due to the addition involved in the total utility function.

In this way, applying the MAUT method to the selection of oil exploration projects has made it possible to put forward a ranking and therefore a portfolio of projects by grouping together the ones that best match the preferences of the decision maker. The main advantages of this method lie in its ability to take into account probabilistic data, to model the preferences/requirements of decision makers in the form of a co-constructed mathematical function, as well as the method's multi-objective approach, which is particularly relevant in the case of a complex decision-making process. On the other hand, the MAUT method was applied in this case with relatively basic initial conditions: partial linear utility functions, additive aggregation, independent criteria, and acceptable compensation. However, this method can take into account much more complex conditions in order to provide a closer representation of the real world. We will note some of the potential improvements in the next section.

4.3. To go further

In this section, we will take a more detailed look at certain aspects of the decision-making process involving the selection of petroleum projects. As mentioned earlier, the case study we considered reflects a relatively basic application of the MAUT method. However, to take into account the complexity of the decision-making problem under consideration, several improvements can be proposed.

4.3.1. *Consideration of the constraints in making the decision*

First of all, the criteria chosen for the selection of the Petrobras project portfolio include the return on investment (NPV), the potential production of hydrocarbons (RES), as well as the external factors (EXT). However, we think it would be pertinent to consider other constraints in making the decision, which act as necessary conditions for the selection of a project. In this sense, the selection of a portfolio of projects is governed by financial, technical and managerial constraints that must be integrated into the decision-making process. This is what the authors Lopes and de Almeida have proposed in their study, by including constraints of a budgetary, geographical and quantitative nature. Classifying projects by total utility will not be sufficient; it is necessary to take into account other differentiating elements that would allow a project to be accepted or rejected.

For this purpose, an algorithm was proposed that sought to maximize the value of the project portfolio (implying the sum of the utilities of the selected projects) without exceeding the available budget (constraint 1), by balancing the geographical areas covered in order to promote broader coverage (constraint 2) and respecting a number of projects defined as relevant (constraint 3). In addition to all this, there is a final constraint which seeks to promote the implications/dependencies between projects (constraint 4): should one project come before another due to technical, managerial or organizational reasons?

Taking into account these operational constraints streamlines the decision by proposing a more realistic selection of projects, which is contextualized and linked to the incompressible and strategic factors specific to these problems in the decision.

4.3.2. *Synergies between criteria/between projects*

In applying the MAUT method proposed in section 4.2, we have implicitly assumed that the decision criteria were uncorrelated and that there was no interdependence. However, this assumption is still questionable, since it is difficult to prove. Is a project with high potential reserves not directly a project with a higher return on investment as a result? Are the external factors impacting an oil exploration project completely independent of the potential exploitation of the available reserves? These are some of the questions that can be raised, reflecting on the one hand the importance of a clear and precise definition of the decision criteria (to avoid redundancies and areas of overlap), and on the other hand the problems with modeling the complexity of a decision-making problem. It is necessary to represent reality as closely as possible, though the use of methods and tools implies a certain degree of standardization and a potential loss of information due to the process modeling approach.

In a later study, Lopes and de Almeida (2015) integrated this notion of dependency, but through the definition of synergies between the selected projects. In this way, the overall utility of a portfolio of projects is considered the sum of the utilities of the projects included in the portfolio, in addition to the synergy effect (positive or negative) existing between them (Keeney and Raiffa 1976; Keeney 1996).

In the context of oil exploration projects, three types of synergies have been identified:

– *Design synergies*: these include, for example, joint developments, shared raw materials and complementary fields of action (in terms of geography or activity). These design synergies allow for economies of scale.

– *Tax synergies*: taxation in the oil industry follows special rules, especially in terms of contracting on the rights to exploit mineral resources. By adequately combining the types of contracts governing the ownership of resources as well as the scope of these contracts, it is possible to create profitable associations at the project portfolio level.

– *Information synergies of information*: the effectiveness of the retrieval and sharing of information, and the establishing of continuity in information from one project to another.

There are a number of ways these synergies can be taken into account. First, the synergistic effect of a project can be considered as an additional decision criterion. The decision model will then have four criteria: NPV, RES, EXT and SYN, represented by a numerical scale from 1 to 5, where 1 means that this project does not presuppose any positive synergy, and 5 means that there is a very high synergy. Projects are then evaluated individually using an index that takes into account their influence on other projects, divided into two groups: those within the territory of Brazil, and those that involve other countries. This assessment is conducted by the international management committee of Petrobras. This means the selection of projects to be included in the portfolio may potentially change through adding an additional criterion to the decision model.

Secondly, synergistic effects between projects can be modelled and measured using computer programs, such as a Monte Carlo simulation. This modeling makes it possible to simulate the portfolios with the highest utility while maximizing the positive synergies between projects. Based on the results that are obtained, taking these synergies into account does not radically change the content of the ideal project portfolio, but it allows a finer level of analysis of the projects and a more realistic vision of how they will be implemented.

4.3.3. *The accuracy of the utility function*

There are improvements that can be proposed regarding the constraints of the problems involved in the decision, on the decision-making model, and therefore on the criteria to be taken into account and the interactions existing between them. Finally, improvements can be proposed on the creation of utility functions. In this sense, we can act on two levels: the aggregation of the utility function and the partial utility function.

In the case presented in section 4.2, we chose to use an additive aggregation of partial utility functions, to obtain the total utility of a project. This additive aggregation is the simplest method, but other methods are also possible, as explained in section 4.2.2.4.

Regarding the construction of partial utility functions (and therefore that of each criterion), the authors proposed to choose a linear function. In a later study, the same authors proposed a different method for constructing these utility functions. They performed an ordinal regression from a scatter plot that was produced from questioning the decision maker directly (direct method). These regressions have made it possible to define more complex and precise utility functions which thus more closely reflect the requirements set by the decision maker on each of the established criteria. Once again, acting with the precision of utility functions makes it possible to propose a decision-making aid that is anchored in reality, that best reflects the complexity of the problem.

4.4. The MAUT method: instructions for use

4.4.1. *MAUT step by step*

To apply the MAUT method, this practical guide offers four steps to follow:

– creation of the alternatives evaluation matrix;

– construction of partial utility functions;

– the weighting of the criteria;

– calculation of total utility and decision.

To illustrate this application step by step, we will apply this method to the choice of a smartphone (inspired by Ishizaka and Nemery 2013).

4.4.1.1. *Alternatives evaluation matrix*

Here, we will consider a problem where we need to rank five fictitious smartphones (SP1–SP5). These five smartphones represent the alternatives that we are looking to separate. The evaluation criteria chosen by the decision maker are as follows:

– the price of the smartphone, given in euros;

– the quality of the post-sales service (PSS), evaluated by a rating between 1 and 5 where 1 is "bad" and 5 is "very good";

– the screen size, measured in inches;

– the storage capacity of the smartphone, rated in GB.

The five smartphones were then evaluated as shown in Table 4.4, using their technical data and opinions of experts.

	Price (€)	PSS	Screen size	Storage capacity
SP1	429	4	4.65	32
SP2	649	4	3.5	64
SP3	459	5	4.3	32
SP4	419	3.5	4.5	16
SP5	519	4.8	4.7	16

Table 4.4. *Evaluation matrix*

4.4.1.2. *Construction of partial utility functions*

For this application, a manual construction method was applied in order to illustrate an applicable case without computer support tools. The decision maker was asked to provide a list of preferences for each of the criteria. This decision maker was then asked to rank the five smartphones for a given criterion.

From this, it became clear that minimizing the price of the smartphone was important to the decision maker, who was also looking for PSS that was as effective as possible, as well as to maximize storage capacity. On the other hand, regarding the screen size, the decision maker seemed to prefer a target value of around 4.5 inches. Based on these observations, a series of linear utility functions were built.

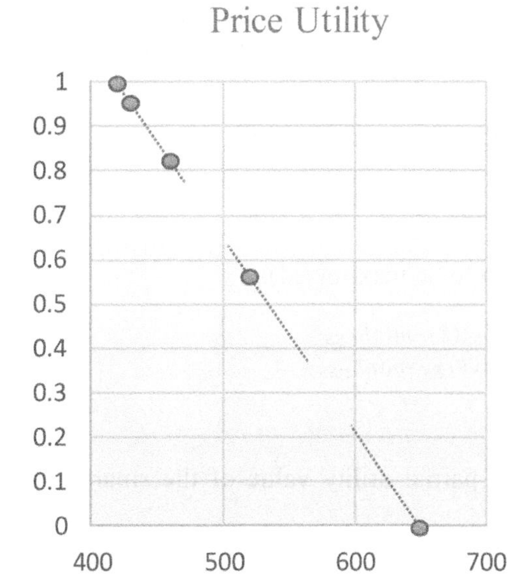

Figure 4.7. *Utility function for the price criterion*

CAUTIONARY NOTE 1.– The utility function provides a lot of information about the profile of the decision maker. It reflects both the goal associated with a criterion (Are we trying to maximize it or minimize it? Do we want to reach a target value?) as well as the requirement level of the decision maker concerning this criterion. This is a useful modeling tool that needs to be as accurate as possible. In our example, the choice was made to focus on linear functions to simplify the calculations while allowing an interesting interpretation of the decision maker's preferences.

For the price (a criterion to be minimized):

$$U_{price}(X) = 1 + \frac{minE_{price} - E_{price}(X)}{maxE_{price} - minE_{price}}$$

where:

– $U_{price}(X)$ is the partial utility value of the smartphone X for the price criterion;

– $E_{price}(X)$ is the evaluation value of the smartphone X for the price criterion;

– $minE_{price}$ is the minimum evaluation value for the price criterion from among the set of smartphones considered (the price of the cheapest smartphone);

– $maxE_{price}$ is the maximum evaluation value for the price criterion from among the set of smartphones considered (the price of the most expensive smartphone).

For PSS (criterion to be maximized):

$$U_{PSS}(X) = \frac{E_{PSS}(X) - minE_{PSS}}{maxE_{PSS} - minE_{PSS}}$$

where:

– $U_{PSS}(X)$ is the partial utility value of the smartphone X for the PSS criterion;

– $E_{PSS}(X)$ is the evaluation value of the smartphone X for the PSS criterion;

– $minE_{PSS}$ is the minimum evaluation value for the PSS criterion, among the set of smartphones considered;

– $maxE_{PSS}$ is the maximum evaluation value for the PSS criterion from among the set of smartphones considered.

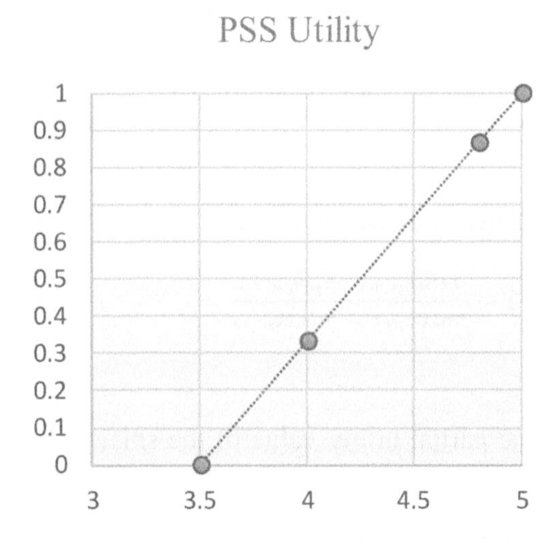

Figure 4.8. *Utility function of the PSS criterion*

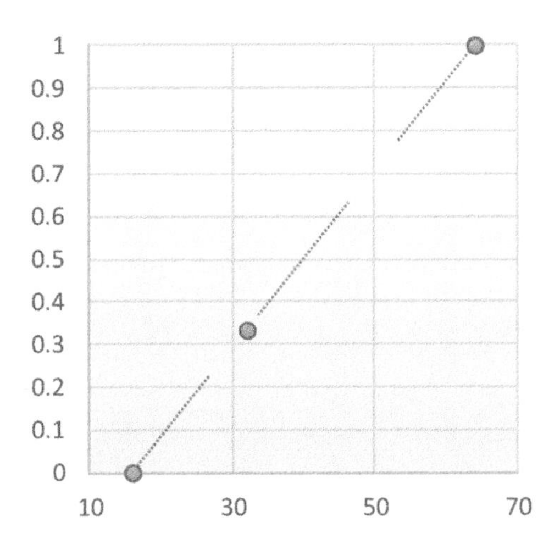

Figure 4.9. *Utility function of the capacity criterion*

For storage capacity (criterion to be maximized):

$$U_{capa}(X) = \frac{E_{capa}(X) - minE_{capa}}{maxE_{capa} - minE_{capa}}$$

where:

– $U_{capa}(X)$ is the partial utility value of the smartphone X for the storage capacity criterion;

– $E_{capa}(X)$ is the evaluation value of the smartphone X for the storage capacity criterion;

– $minE_{capa}$ is the minimum evaluation value for the storage capacity criterion, from among the set of smartphones considered;

– $maxE_{capa}$ is the maximum evaluation value for the storage capacity criterion, from among the set of smartphones considered;

For the screen size (target value = 4.5), see Figure 4.10.

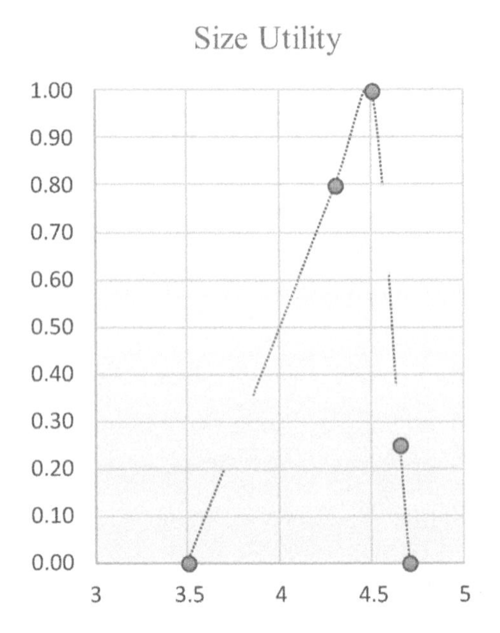

Figure 4.10. *Utility function of the size criterion*

For $E_{size}(X) < 4.5$:

$$U_{size}(X) = \frac{E_{size}(X) - minE_{size}}{4.5 - minE_{size}}$$

For $E_{size}(X) > 4.5$:

$$U_{size}(X) = 1 + \frac{4.5 - E_{size}(X)}{maxE_{size} - 4.5}$$

where:

– $U_{size}(X)$ is the partial utility value of the smartphone X for the screen size criterion;

– $E_{size}(X)$ is the rating of the smartphone X for the screen size criterion;

– $minE_{size}$ and $maxE_{size}$ are the minimum/maximum evaluation values for the screen size criterion from among the set of smartphones considered.

4.4.1.3. *Weighting of the criteria*

The decision maker was then asked to assign a relative importance to the evaluation criteria that had been chosen. This would be done by distributing one hundred tokens among the four criteria, so as to represent the importance of these criteria in making the decision.

This process was used to obtain the weights presented in Table 4.5.

	Price	PSS	Screen size	Storage capacity
Weighting	35%	35%	15%	15%

Table 4.5. *Weighting of criteria*

For the decision maker, the most important criteria are the price of the smartphone and the quality of its PSS (which each received 35%). The screen size and storage capacity are shown to be secondary, with a weighting of 15% each.

CAUTIONARY NOTE 2.– To construct the weights of the criteria, the AHP method could also have been used (see Chapter 2). The decision maker would then have made pairwise comparisons of each criterion to determine their relative importance.

4.4.1.4. *Calculation of total utility and decision*

The final step in applying MAUT is to aggregate the partial utilities into a utility value total, reflecting the overall "desirability" of an alternative for the decision maker.

There are several aggregations that can be conceived of. For the sake of simplicity for this example, we will choose an additive aggregation. In this way, we will calculate the total utility of each smartphone using the following formula:

$$U_{TOT}(X) = \sum_{i=1}^{4} \omega_i \cdot U_i(X)$$

where:

– $U_{TOT}(X)$ is the total utility of the smartphone X;

– ω_i is the weighting of criterion *i*;

– $U_i(X)$ is the partial utility of the smartphone X for criterion *i*.

CAUTIONARY NOTE 3.– The choice of additive aggregation is particularly appropriate in the case where the criteria are not correlated with each other, which is not quite the case here. Indeed, the size of the screen, the storage capacity, and the PSS seem to be independent, while on the other hand, the price is certainly related to the technical specificities of smartphones, particularly the storage capacity. Thus, the choice of additive aggregation is questionable, but it has the advantage that it can be applied using simple mathematical calculations that do not require digital support tools.

The results of these calculations are shown in Figure 4.11.

Smartphones	Total Utility
SP1	0.267
SP2	0.456
SP3	0.809
SP4	0.500
SP5	0.501

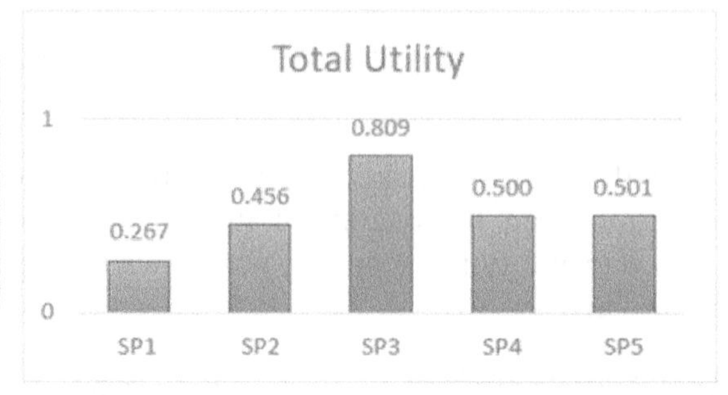

Figure 4.11. *Ranking of smartphones*

The smartphone that appears the best suited to the preferences of the decision maker would seem to be smartphone 3, which has a significant lead. After that come smartphones 2 4, and 5, which are very close together, then finally smartphone 1, which appears as the least suitable for the decision maker.

This illustration of using the MAUT method on a simple application case allows us to highlight the importance of the decision maker's involvement in the application of a method like this one. Indeed, all of the data necessary for the calculation is given directly by the decision maker: the criteria to be taken into account, the partial utility functions, the weightings, etc. The MAUT method thus makes it possible to mathematically translate a preference and re-transcribe so as to arrive at a model for supporting decisions. Although this example has been deliberately simplified, these steps represent the traditional process of solving a decision-making problem with the MAUT method.

4.4.2. *Using software supports*

For greater accuracy in the application of MAUT, it is necessary to use digital support tools almost systematically to manage the complexity that arises due to modeling, simulation, algorithms, etc. The MAUT method, based on initially basic computational steps, can thus benefit from support of computer software, allowing for more extensive applications. This makes it possible to improve the accuracy of partial utility functions, the weightings of the criteria, or for choosing a more realistic approval method of the total utility.

DecideIT[2], D-Sight[3] and Logical Decision[4] are all software programs that implement the MAUT method and are distributed in the form of licenses. We will also look at the Right Choice software[5], developed by Ventana System UK and available in Open Access.

In general, these programs include different features, such as:

– modeling the decision-making problem as a multi-level tree of alternatives and criteria for defining groups and subgroups;

– the defining of the importance of each criterion using a weighting tool;

– the assignment of different types of utility functions (linear, exponential, logarithmic, echelon, quadratic, etc.) and the ability to adjust their parameters;

– the ranking of alternatives according to their total utility value total;

– access to a sensitivity analysis, allowing the weighting values of the criteria to be changed and to instantly view the changes this creates on the ranking of the alternatives;

– the option to add different scenarios in terms of preference parameters, which allows the application of this software for group decision making.

Thus, many research works rely on these software programs to offer more detailed analyses of the decision-making issues they address. For example, Min (1994) proposes an application of MAUT using the Logical Decision

2 Available at: www.preference.nu/decideit/.

3 Available at: www.d-sight.com/.

4 Available at: www.logicaldecisions.com/.

5 Available at: www.ventanasystems.co.uk/services/software/rightchoice/.

software for optimizing the selection of international suppliers in the IT sector. It is based on a reference set of seven selection criteria, divided into different attributes, and it evaluates five potential suppliers according to this decision-making model. Once the weighting process is complete, as well as the construction of utility functions associated with the criteria, it compares the ranking of suppliers obtained with a sensitivity analysis, allowing us to set out different scenarios in terms of weighting. While there are some cases where one or two suppliers clearly stand out from the others on a given criterion, in other cases the ranking is more sensitive and can vary greatly if a small adjustment is made to the weightings given to the criteria. These sensitivity analyses shed more light on the decision-making process, with the software allowing for greater nuance in the ranking. The results should not be taken as an absolute truth, but as a representation of decision assistance to be studied and discussed by the decision maker.

In order to provide an illustration of the benefits of using software programs for applying MAUT, let us return to our previous example and analyze it using the program Right Choice.

We will consider five smartphones to be ranked according to the preferences of a decision maker. The criteria are the same as before, and the weights assigned to them also remain unchanged (Figure 4.12).

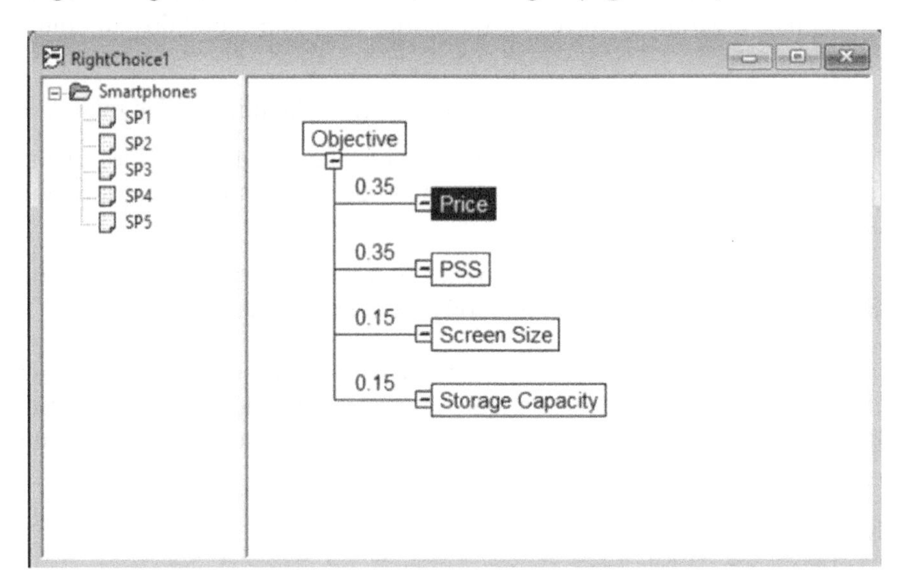

Figure 4.12. *Architecture of the decision problem – Right Choice*

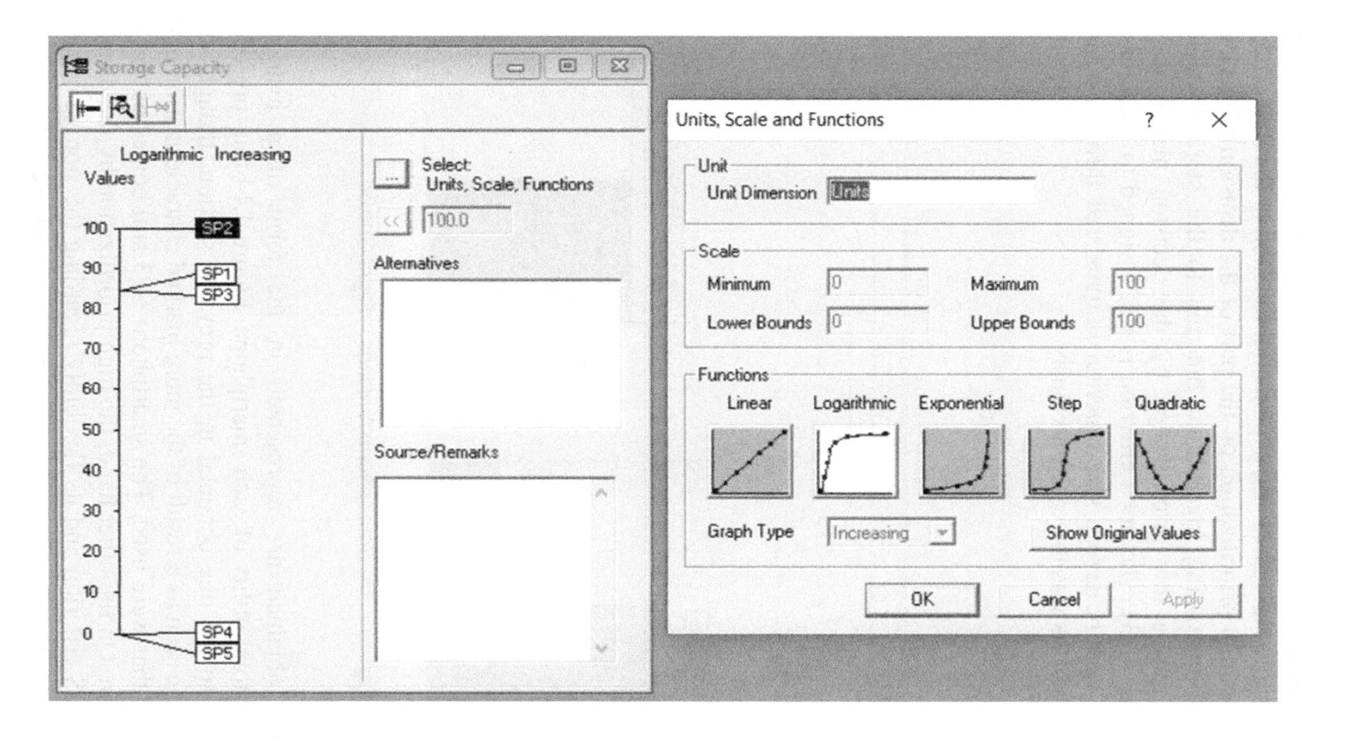

Figure 4.13. *Assignment of utility functions – Right Choice*

One of the main advantages of using Right Choice is that it makes it possible to consider the use of more complex utility functions. So, for this example, we assigned an increasing exponential function to the PSS criterion (which reflects a high prioritization on the part of the decision maker), and we assigned an increasing logarithmic function to the storage capacity criterion (reflecting a low prioritization by the decision maker on this criterion) (Figure 4.13). The price criterion is still assigned to a decreasing linear function, and the screen size criterion is translated through a target value to be reached, as was the case previously.

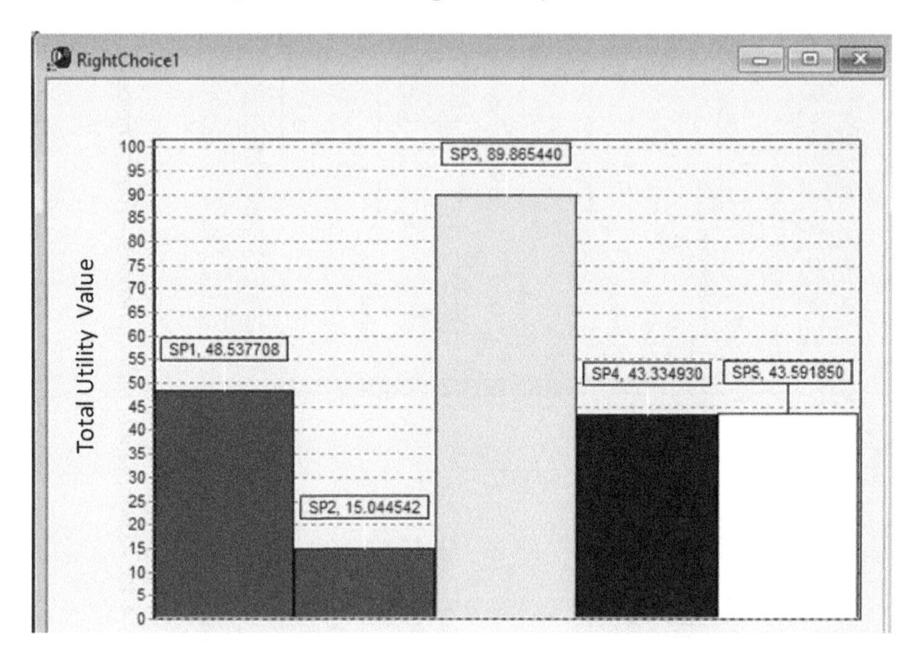

Figure 4.14. *Smartphone rankings – Right Choice*

The results obtained are represented in the form of a histogram highlighting the total utility of each smartphone. Figure 4.14 thus presents a ranking similar to the one obtained in the previous section, though some differences can be seen as a result of the more marked characterization of the utility functions that were used. The smartphone SP3 still comes out on top by a wide margin. On the other hand, the high prioritization by the decision maker of the PSS criterion and its flexibility regarding the storage capacity have a negative impact on the total utility of the smartphone SP2, which is thus ranked last. On the other hand, smartphone SP1 is positively impacted by these changes in utility functions: it has moved up in the ranking to second place.

Finally, a sensitivity analysis can be carried out to highlight the changes in the ranking of smartphones according to the weights assigned to the criteria. Figure 4.15 shows the change in the ranking according to the value defined for the weightings of the screen size criterion. While the smartphone SP3 still remains in the lead, we can see the changes that have occurred with other smartphones. In our example, these sensitivity analyses do not have many practical implementations since the smartphone SP3 strongly dominates the other alternatives, but this feature of the software can be particularly useful in the case where the ranking is not so clearly defined (see section 6.3).

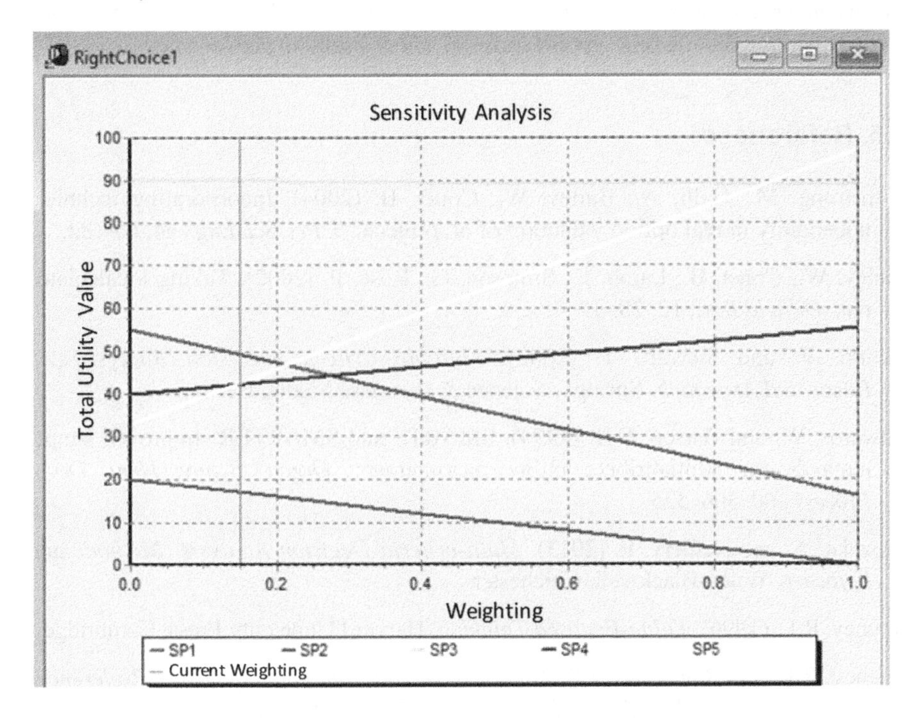

Figure 4.15. *Sensitivity analysis – Right Choice. For a color version of this figure, see www.iste.co.uk/enjolras/decisionmaking.zip*

MAUT AT A GLANCE.– Objective: MAUT is a ranking method seeking to rank alternatives according to an order corresponding to the preferences of the decision maker.

Unique feature: it is based on the notion of utility, which allows it to model the requirements of the decision maker according to the criteria taken into account in the issues related to the decision. This utility is then translated through a mathematical function representing the satisfaction of the decision maker with regard to the performance achieved by the alternatives considered.

Limitations: the computational principle behind the MAUT method is the weighted average. This means that it is a compensatory method, where a very poor evaluation on one criterion can be made up for by a strong performance on another criterion. In addition, the construction of the utility function associated with the decision criteria is a crucial step, but it is sometimes difficult to implement operationally. The use of a support software is thus useful to gain accuracy.

4.5. References

Armstrong, M., Galli, A., Bailey, W., Couët, B. (2004). Incorporating technical uncertainty in real option valuation of oil projects. *J. Pet. Sci. Eng.*, 44, 67– 82.

Bailey, W., Couet, B., Lamb, F., Simpson, G., Rose, P. (2000). Taking a calculated risk. *Oilfield Rev.*, 12, 20–35.

Belton, V. and Stewart, T. (2002). *Multiple Criteria Decision Analysis: An Integrated Approach*. Springer Science & Business Media, Berlin.

Edwards, W. and Barron, F.H. (1994). SMARTS and SMARTER: Improved simple methods for multiattribute utility measurement. *Organ. Behav. Hum. Decis. Process.*, 60, 306–325.

Ishizaka, A. and Nemery, P. (2013). *Multi-criteria Decision Analysis: Methods and Software*. Wiley-Blackwell, Chichester.

Keeney, R.L. (1996). *Value-Focused Thinking*. Harvard University Press, Cambridge.

Keeney, R.L. and Raiffa, H. (1993). *Decisions with Multiple Objectives: Preferences and Value Trade-Offs*. Cambridge University Press, Cambridge.

Lopes, Y.G. and de Almeida, A.T. (2013). A multicriteria decision model for selecting a portfolio of oil and gas exploration projects. *Pesqui. Oper.*, 33, 417–441.

Lopes, Y.G. and de Almeida, A.T. (2015). Assessment of synergies for selecting a project portfolio in the petroleum industry based on a multi-attribute utility function. *J. Pet. Sci. Eng.*, 126, 131–140.

Min, H. (1994). International supplier selection: A multi-attribute utility approach. *Int. J. Phys. Distrib. Logist. Manag.*, 24, 24–33.

Neumann, J.V. and Morgenstern, O. (1953). *Theory of Games and Economic Behavior*. Princeton University Press, Princeton.

Zionts, S. (1992). Some thoughts on research in multiple criteria decision making. *Comput. Oper. Res.*, 19, 567–570.

5

The Recruitment Process in Human Resources: An Application of ELECTRE

Human resources are part of the capital of a company. Indeed, the human factor is considered an important resource in an organization (Liu et al. 2012). Skill sets, know-how and knowledge are thus an intangible asset for a company. It is therefore necessary to put in place a system to manage human resources, in order to properly respond to the needs of the development of current projects, but also by adopting a forward-looking vision to define the future skill sets to be acquired. To accomplish this, companies have different options: training current staff to integrate key skills, recruitment of new employees trained in these skills or implementing external talent in the form of services or collaborations with ecosystem actors (open source, partnerships, communities, etc.) (Altman et al. 2014).

In this chapter, we will focus on the recruitment process of new staff members. Personnel selection is primarily an intuitive practice. The recruitment managers analyze the candidates' profiles, conduct interviews and finally make a decision based on their expertise. The choice of the most suitable candidate traditionally depends on several factors: academic factors (such as level of study), technical factors (such as mastery of specific software), professional factors (such as experience) and personal factors (such as inter-personal skills). All of these factors must be considered in

Decision-Making Tools to Support Innovation,
by Manon ENJOLRAS, Daniel GALVEZ and Mauricio CAMARGO. © ISTE Ltd 2023.

order to carry out a detailed analysis. The qualitative nature and diversity of these factors mean that making a decision on new hires is particularly complex.

In order to reduce uncertainty and subjectivity in choosing new staff, various authors have proposed multi-criteria approaches as tools to support this process. We will analyze the case put forward by Demirci and Kilic (2019), who implement three methods to assist in the selection of staff members in a manufacturing company. The three methods used are Dematel, ANP and ELECTRE. While Dematel seeks to establish the relationships of influence between the criteria (see Chapter 2), ANP helps to establish the importance of each criterion for making the decisions (using a model similar to that of the method it is based on: AHP). Finally, ELECTRE allows for a comparison of candidates in order to choose the most suitable person for the position offered. We will study the application of ELECTRE and the contributions related to its use to prioritize candidates in the context of personnel selection.

5.1. Context and challenges in decision-making

5.1.1. *Human resources management and innovation*

The knowledge, skills and abilities of people working within a company are an integral part of the development of its capacity for innovation and absorption (Lenihan et al. 2019). In the 1970s, most companies created value by producing tangible assets. Today, by contrast, most companies create value from intellectual property rights and the top-tier qualifications of their human resources (Bircan and Gençler 2015). Various studies have shown a positive relationship between intellectual capital and company performance (Chen Goh 2005). Intellectual and human capital supports creativity, risk-taking and problem-solving by directly impacting the efficiency and performance of the organization. Human capital includes different attributes, such as personality, creativity, well-being, self-motivation and resilience (Madrid et al. 2018). The role of motivation at work and the variety of tasks assigned to staff members is crucial to ensuring the innovation and success of organizations (Chiu 2018). If we are interested in the relationship between human capital and innovation, several elements are to be considered as supports (Lenihan et al. 2019): job satisfaction, a propensity for change, a sense of belonging and the search for compromise with the organization. Employees who are satisfied with their work are likely to put in extra effort, take more risks, learn new skills and generate new ideas. The propensity to change can be linked to the attraction for new technologies, new skills or new responsibilities. Thus, a mentality that favors change impacts how new innovations are adopted. The feeling of belonging to the organization allows employees to feel confident enough to step up and take risks, propose new ideas and get involved to help with the success of the company. The company's human resources management system must therefore encourage these key elements of success by promoting job satisfaction and a sense of belonging, while motivating its employees' willingness to take risks and make changes (Lenihan et al. 2019). To do this, the human resources management system oversees the training and monitoring activities of the staff members who work at the organization. However, there is another key activity done by the human resources management system: recruitment.

5.1.2. *The challenges of the recruitment process*

Choosing who to hire is a process that all companies go through. The objective is to choose the most suitable person for a position to be filled from among several different candidates. The recruitment process consists of different stages, the most frequently occurring of which are (1) the application, usually done via a file presenting the candidates' personal, academic and professional information (a curriculum vitae) as well as a description of the factors motivating them to apply; (2) interviews with the recruitment manager to get to know and expand on certain details (sometimes supplemented with psychological tests); and (3) the selection stage. This step in particular is often critical for companies. Most of them make the decision intuitively based on the experience and vision of the head of recruiting operations. The lack of a standardized process or model for supporting new staff hires has thus garnered the interest of various different authors.

Recruitment depends on several factors; university education, professional experience, technical skills, language skills and personal aptitudes are all criteria that have traditionally been analyzed in choosing staff. The many different factors involved have given rise to multi-criteria approaches to support the recruitment process. Table 5.1 highlights the diversity of factors involved in choosing new staff. Although some are recurring, the factors to be evaluated are directly related to the job profile defined by the company, according to its objectives and needs. Most of the factors have a general qualitative assessment associated with a linguistic scale (e.g. in the form of "bad", "average", "good", "very good").

The challenge is thus to identify the relevant factors or criteria associated with an adequate evaluation scale in order to be able to compare candidates and ensure that the right profile is found for the position to be filled. The definition of the criteria, as well as their evaluation scale, will be the foundations for building a model for supporting hiring decisions based on multi-criteria methods.

Author	Criterion	Evaluation	Method
(Sang et al. 2015)	– Emotional stability – Verbal communication – Personality – Experience – Self-confidence	Linguistic scale (with seven values)	Fuzzy TOPSIS
(Dursun and Karsak 2010)	– Emotional stability – Leadership – Self-confidence – Verbal communication – Personality – Experience – General aptitude – Understanding	Linguistic scale used for the first six criteria. The last two criteria use a point-based scale (0–100)	OWA and fuzzy TOPSIS
(Güngör et al. 2009)	Three categories of criteria: – General factors (experience, foreign language, academic qualifications, knowledge of integrated systems and computer science) – Complementary factors (decision-making, teamwork, effective time management, determination, learning, motivation) – Individual factors (basic skills, appearance, age, culture and communication)	Linguistic scale	Fuzzy AHP
(Ijadi Maghsoodi et al. 2020)	– Degrees – Academic performance – Knowledge – Psychological factors – General factors	Linguistic scale	Multi-objective

	– Mobility – Salary expectations – Flexibility		
(Demirci and Kilic 2019)	– Education – Experience – Personality and personal skills – Technical skills – Foreign languages – Mobility – Results of the examination done by the company	Point scale (0–100)	Dematel, ANP, and ELECTRE

Table 5.1. *Analysis of multi-criteria approaches as part of the hiring process*

5.2. The ELECTRE method

The "Elimination and Choice Expressing Reality"[1] (ELECTRE) method was created by Bernard Roy in the late 1960s (Roy 1968). ELECTRE was one of the first "outranking" methods developed, which further inspired several European works on multi-criteria analysis. Outranking methods are based on dominance relationships between the alternatives.

The ELECTRE method offers different variants to be used depending on the objectives of different decision makers. The first version is the ELECTRE-I, which lays the foundations for the way the method operates. Figure 5.1 illustrates the principal steps and concepts used in the ELECTRE method.

5.2.1. *Methodological concepts*

The input data for the ELECTRE application are as follows:

– The evaluation of alternatives in relation to criteria: like all discrete multi-criteria methods, ELECTRE works on the prioritization of previously defined alternatives. Each alternative must be evaluated using a list of selection criteria which allow decision makers to differentiate and order the alternatives.

– The weight vector: each criterion is characterized by its importance in making the decision. This importance is represented by a percentage, and the sum of the importance must be 100%.

1 ELECTRE - ÉLimination et Choix Traduisant la REalité.

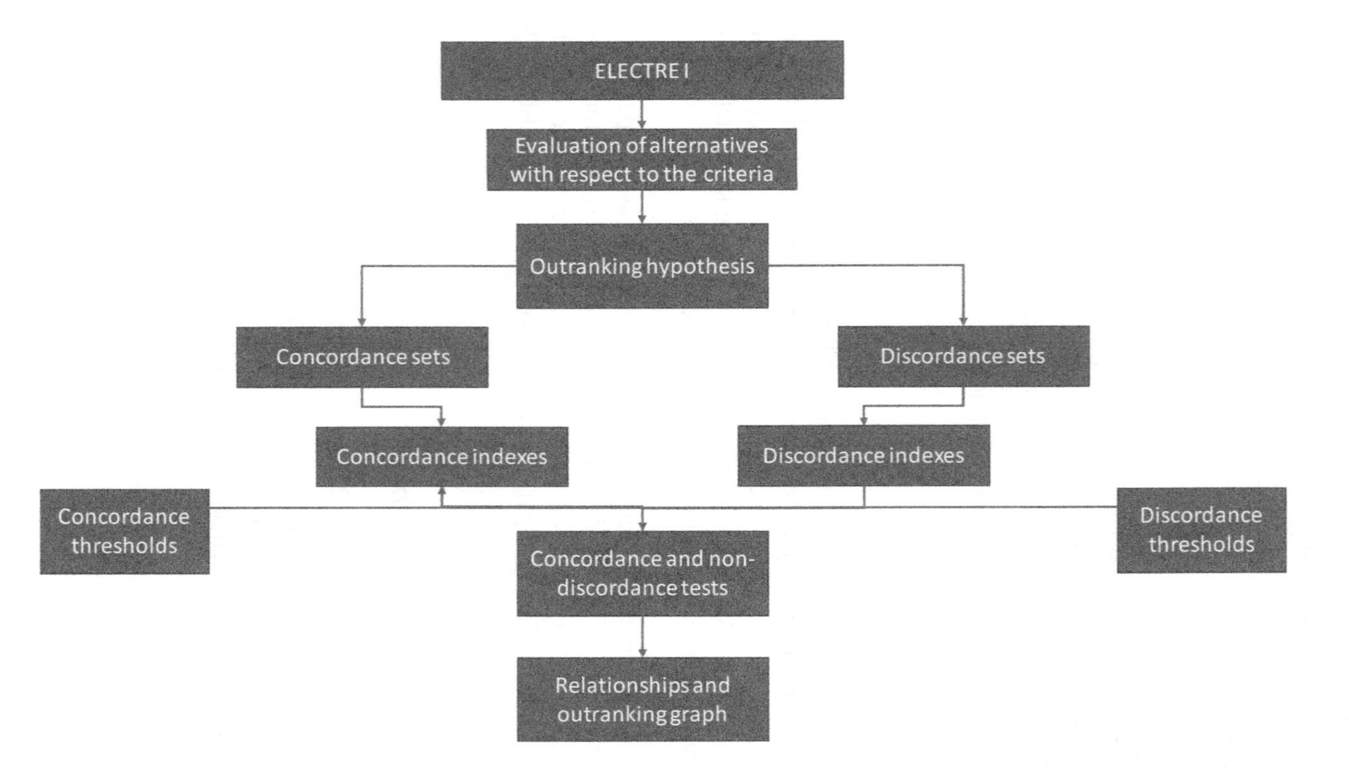

Figure 5.1. *Explanation of the ELECTRE-I method (Maystre et al. 1994)*

Based on these data, the ELECTRE method can be then applied. The objective of the method is to obtain all the information necessary to validate and not refute the outranking hypothesis from one alternative to another alternative. To validate the outranking hypothesis, it is thus necessary to collect all information that ensures that an alternative is better evaluated than another alternative. In ELECTRE, this information is defined as the "concordance". The concordance index is between 0 and 1 and represents the percentage of information collected to validate the outranking hypothesis.

On the other hand, the lack of any refutation of the outranking options of the hypothesis is related to the information that ensures that an alternative is evaluated less favorably than another alternative. This information is reflected by the concept of discordance. The discordance index is also between 0 and 1 and represents the percentage of information collected to refute the outranking hypothesis.

Then, it is necessary to establish the minimum percentage of information to validate the outranking hypothesis; that is, it is necessary to define a concordance threshold. All concordance indexes equal to or greater than this threshold will reflect a sufficient amount of information to validate the hypothesis. On the other hand, it is necessary to establish the maximum percentage of information that will be tolerated in order not to refute the outranking hypothesis, and therefore set a discordance threshold. Any discordance indexes equal to or less than this threshold have not reached the amount of information sufficient to refute the hypothesis.

Since the objective of ELECTRE is to both validate and not refute the hypothesis, only pairwise comparisons that have a concordance index equal to or greater than the concordance threshold and, at the same time, an discordance index equal to or less than the discordance threshold, will ensure that the first alternative of the comparison outperforms the second. We may refer to this as "dominance" by the first alternative over the second alternative. These dominances between alternatives are indicated on a graph that compares the alternatives by taking an arrow out of the dominant alternative and arriving at the dominated alternative. This graph is produced by ELECTRE, allowing the alternatives to be prioritized by finding the alternative that dominates the others.

There are several limitations to applying the ELECTRE-I method. For example, if the set thresholds are very demanding, that is, with a high concordance index (more than 50%) and a low discordance index (less than 50%), the dominance relationships will be very restricted. In this case, it may occur that one or more alternatives do not show any dominant relationship over the others. It is then impossible to establish a complete ranking because these alternatives will fall within a zone of incomparability.

In order to allow the method to progress and to partially make up for its limitations, various extended versions of ELECTRE have been developed. Roy (1976) has classified these versions according to the type of problem they solve:

– the ELECTRE-I and ELECTRE-IS methods are used for problems involving choices where the objective is to select an alternative or a small group of alternatives that best meet the preferences of decision makers;

– the ELECTRE-II, ELECTRE-III and ELECTRE-IV methods were created to be used for scheduling or ranking problems on larger groups;

– the ELECTRE-TRI method seeks to solve *sorting* problems where the alternatives will be ordered but will also be classified into previously defined categories (Govindan and Jepsen 2016).

In this chapter, we will study the ELECTRE-I version.

5.2.2. *Application of ELECTRE to the hiring of new staff members*

This section will study the application of the ELECTRE method within the hiring process, based on the study by Demirci and Kilic (2019). These authors study the selection of new staff members for a company in the manufacturing industry that would like to recruit engineers. The company uses a recruitment process that is based on three stages (Figure 5.2): first, candidates are contacted for a telephone interview to assess their personality, foreign language proficiency and technical skills. Second, a pre-selection of candidates is carried out based on this interview. Finally, the successful candidates are received by the director in charge of the department where the position will be based. This director is responsible for the final decision regarding the candidate to be hired, and today this decision is completely intuitive.

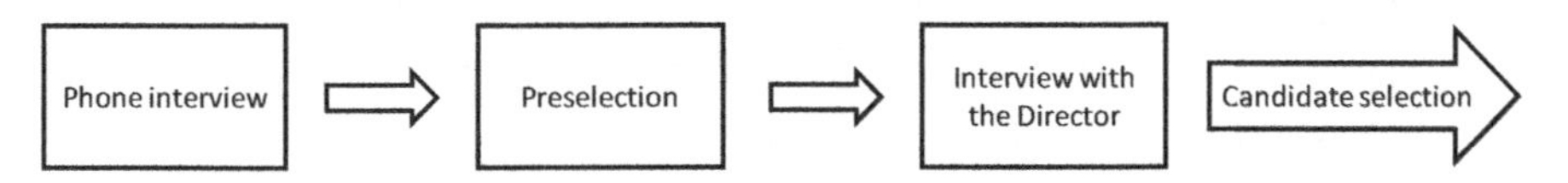

Figure 5.2. *Staff selection process (Demirci and Kilic 2019)*

The approach proposed by Demirci and Kilic (2019) seeks to improve the third stage of the process: the interview with the director. The first two steps remain unchanged, and the third will be supported by the application of the ELECTRE method to add objectivity to the final decision.

Seven criteria are then defined based on the opinion of experts in human resources management inside company, taking into account a review of the academic literature:

– education: a Bachelor's degree in industrial, mechanical, mechatronic or materials engineering is defined as a minimum requirement;

– experience: experience in similar jobs;

– personality and personal skills: this criterion is assessed directly during telephone interviews and with the director;

– technical skills: the key technical skills are related to the knowledge of continuous improvement methods (Kaizen, 5s, Lean management);

foreign language: in this case, English and/or German are the languages desired;

– mobility: availability to travel abroad;

– results on the exam made by the company: a score between 0 and 100 obtained on the test.

Seven candidates have been shortlisted to reach this third stage. They will then be evaluated using the multi-criteria approach proposed in the study by Demirci and Kilic. These authors propose a combination of three multi-criteria methods to bolster the hiring process used in their case study (Figure 5.3): Dematel, ANP and ELECTRE.

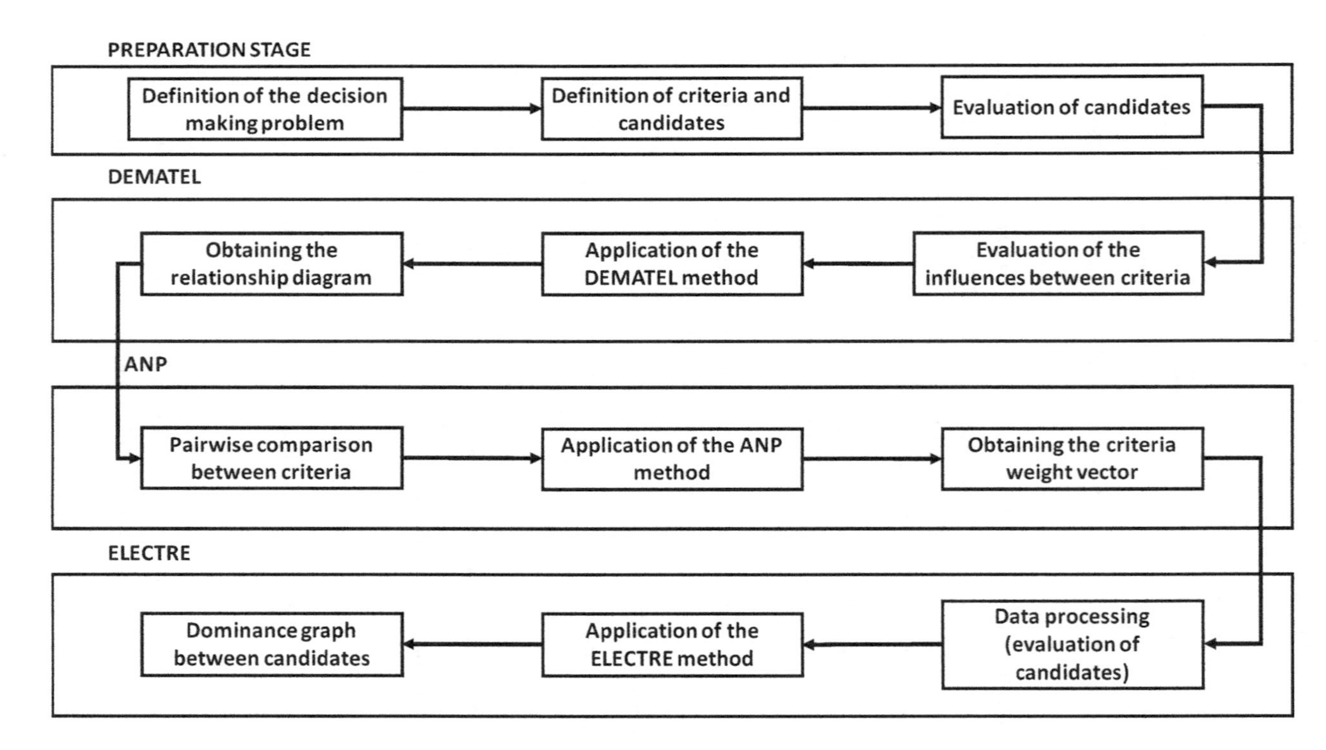

Figure 5.3. *Approach proposed by Demirci and Kilic (2019)*

5.2.2.1. *Preliminary steps*

Once the context of decision-making has been identified and the decision-making model has been established (i.e. the identification of the criteria considered and the candidates to be evaluated), the first step is to use the DEMATEL method in order to find the relationships between the seven criteria described above. The result of this step is a diagram highlighting the existing interrelationships between the criteria (the application of the DEMATEL method is detailed in Chapter 2). Then, an application of the ANP method allows us to define the importance of each criterion for selecting candidates. In particular, ANP makes it possible to calculate the weighting of each criterion, while taking into account the existing interrelationships between these criteria (ANP is a variation of the AHP method and shares its theoretical foundations; see Chapter 2). The weightings of the criteria are one of the input data needed to apply the ELECTRE method. The set of necessary data is composed of the criteria and the weightings assigned to them, respectively, and the selected alternatives (candidates shortlist) as well as their evaluations according to the defined criteria. All candidates were evaluated by the department manager using a scale from 1 to 100. For all criteria, 100 is the maximum rating. Therefore, all criteria are to be maximized. This information is presented in Table 5.2.

	Education	Experience	Personality	Technical skills	Foreign languages	Mobility	Exam results
C1	100	100	100	80	85	100	100
C2	97	100	80	80	85	100	100
C3	97	100	80	70	75	100	95
C4	97	100	90	85	80	100	80
C5	100	100	80	70	65	100	10
C6	100	100	80	75	70	100	15
C7	100	100	50	70	50	100	20
Weighting	0.293	0.201	0.166	0.149	0.055	0.034	0.099

Table 5.2. *Evaluations of alternatives with respect to the criteria (Demirci and Kilic 2019)*

5.2.2.2. *Application steps of the ELECTRE method*

The first step in the application of ELECTRE is the normalization of the alternative evaluation table (Table 5.2). In this case, the calculation of the normalization is based on the following equation:

$$X_{ij} = \frac{a_{ij}}{\sqrt[2]{\sum_{i=1}^{7} a^2{}_{ij}}}$$

Each evaluation (a_{ij}) is divided by the square root of the sum of all evaluations squared $(a^2{}_{ij})$ of the alternatives for this criterion. For example, for the assessment of the candidate's education criterion 1:

$$X_{11} = 100 \Big/ \sqrt[2]{100^2 + 97^2 + 97^2 + 97^2 + 100^2 + 100^2 + 100^2)} = 0.383.$$

Since all the criteria are to be maximized, the same equation is used for each of them.

Table 5.3 shows the normalization of all evaluations.

	Education	Experience	Personality	Technical skills	Foreign languages	Mobility	Exam results
C1	0.383	0.378	0.465	0.398	0.435	0.378	0.526
C2	0.371	0.378	0.372	0.398	0.435	0.378	0.526
C3	0.371	0.378	0.372	0.348	0.384	0.378	0.500
C4	0.371	0.378	0.419	0.423	0.410	0.378	0.421
C5	0.383	0.378	0.372	0.348	0.333	0.378	0.053
C6	0.383	0.378	0.372	0.373	0.359	0.378	0.079
C7	0.383	0.378	0.233	0.348	0.256	0.378	0.105

Table 5.3. *Normalization of evaluations (Demirci and Kilic 2019)*

The normalized values must then be weighted by multiplying them by the weight corresponding to the criterion. All ratings in the "education" column are thus multiplied by the weighting of this criterion. The ratings in the next column are multiplied by the weight of the criterion "experience", and so on. Table 5.4 shows the standardized and weighted assessments.

	Education	Experience	Personality	Technical skills	Foreign languages	Mobility	Exam results
C1	0.112	0.076	0.078	0.060	0.024	0.013	0.052
C2	0.109	0.076	0.062	0.060	0.024	0.013	0.052
C3	0.109	0.076	0.062	0.052	0.021	0.013	0.049
C4	0.109	0.076	0.070	0.063	0.023	0.013	0.042
C5	0.112	0.076	0.062	0.052	0.018	0.013	0.005
C6	0.112	0.076	0.062	0.056	0.020	0.013	0.008
C7	0.112	0.076	0.039	0.052	0.014	0.013	0.010

Table 5.4. *Normalized and weighted values of evaluations (Demirci and Kilic 2019)*

This table is used to calculate the concordance and discordance indexes.

Concordance represents the information that shows that one alternative is better than another alternative for a given criterion. Then, each alternative must be compared with all the other alternatives, criterion by criterion. If the first alternative (of the comparison) is equal to or better than the second alternative, the weighting of this criterion is added in the concordance index of this comparison. The concordance index of a given comparison is therefore the sum of the weights of the criteria for which the first alternative (of the comparison) is equal to or better than the second alternative. For example, the concordance index of the comparison candidate 1 versus candidate 4: C (1,4) is obtained by the following calculation:

$$C\ (1,4) = 0.293 + 0.201 + 0.166 + 0 + 0.055 + 0.034 + 0.099 = 0.850.$$

Candidate 1 is evaluated only on the criterion “technical skills”, so the concordance index of comparison 1 versus 4 considers all weights except the weight of the criterion “technical skills”.

If we look at the opposite concordance C (4,1):

$$C\ (4,1) = 0 + 0.201 + 0 + 0.149 + 0 + 0.034 + 0 = 0.385.$$

Once the completeness of the concordance indexes are calculated, a concordance matrix can be created:

$$\begin{pmatrix} - & 1.00 & 1.00 & 0.85 & 1.00 & 1.00 & 1.00 \\ 0.54 & - & 1.00 & 0.68 & 0.70 & 0.70 & 0.70 \\ 0.23 & 0.69 & - & 0.62 & 0.70 & 0.55 & 0.70 \\ 0.38 & 0.84 & 0.90 & - & 0.70 & 0.70 & 0.70 \\ 0.52 & 0.69 & 0.84 & 0.52 & - & 0.69 & 0.90 \\ 0.52 & 0.69 & 0.84 & 0.52 & 1.00 & - & 0.90 \\ 0.52 & 0.52 & 0.67 & 0.52 & 0.77 & 0.62 & - \end{pmatrix}.$$

The matrix that is obtained is a square matrix for which the number of rows and columns represents the number of alternatives considered. It is read line by line. In other words, the concordance index for the comparison (A/B) is obtained by considering the alternative of row A with respect to the alternative of column B. These indexes will fall between 0 and 1, since they cannot exceed 100% of the sum of all weights.

The discordance index represents the information that shows that one alternative is equal to or less favorable than another alternative (the opposite result of concordance). Then, a comparison must be made of one alternative with respect to all the others, in order to identify the criteria for which the first alternative (of the comparison) is equal to or worse than the second. For example, for the comparison of candidate 1 with candidate 4, it can be seen from Table 5.4 that candidate 1 is evaluated equally with candidate 4 in the criteria "experience" and "mobility", and he/she is not as good in "technical skills". These three criteria highlight different degrees of discordance. Next, to define the discordance index, the largest difference between the criteria showing discordances should be divided by the maximum difference between the two candidates.

The only criterion for which candidate 1 is worse than candidate 4 is the "technical skills" criterion, while the other discordant criteria have been evaluated equally. The "technical skills" criterion is therefore the one that reflects the maximum degree of discordance. The numerator of the discordance index thus represents the difference between the two candidates for this criterion. Next, the denominator is obtained by identifying the criterion for which the candidates show the greatest difference. This is the criterion "exam results":

$$\mathrm{D(1,4)} = \frac{\max|differences\ between\ criteria\ with\ a\ level\ of\ discordance|}{\max|differences\ between\ all\ criteria|}$$

$$\mathrm{D(1,4)} = \frac{|difference\ technicalskillson|}{|difference\ exam\ results|} = \frac{|0.063-0.060|}{|0.052-0.042|} = 0.3.$$

In this way, the discordance index is the ratio between the maximum level of discordance and the maximum difference between the alternatives. If we look at the discordance of the opposite comparison D(4,1), the criteria that show discordances are education, experience, personality, foreign language, mobility and exam results. Since the maximum difference between the two alternatives appears on a criterion that is also a discordance criterion, for comparison 4.1, the discordance index will then be equal to 1:

$$D(4,1) = \frac{|\text{difference exam results}|}{|\text{difference exam results}|} = \frac{|0.052\text{-}0.042|}{|0.052\text{-}0.042|} = 1.$$

As with concordance, a discordance matrix can thus be constructed:

$$\begin{pmatrix} - & 0.00 & 0.00 & 0.35 & 0.00 & 0.00 & 0.00 \\ 1.00 & - & 0.00 & 0.74 & 0.07 & 0.07 & 0.08 \\ 1.00 & 1.00 & - & 1.00 & 0.07 & 0.09 & 0.08 \\ 1.00 & 1.00 & 0.69 & - & 0.09 & 0.10 & 0.10 \\ 1.00 & 1.00 & 1.00 & 1.00 & - & 1.00 & 0.22 \\ 1.00 & 1.00 & 1.00 & 1.00 & 0.00 & - & 0.11 \\ 1.00 & 1.00 & 1.00 & 1.00 & 1.00 & 1.00 & - \end{pmatrix}.$$

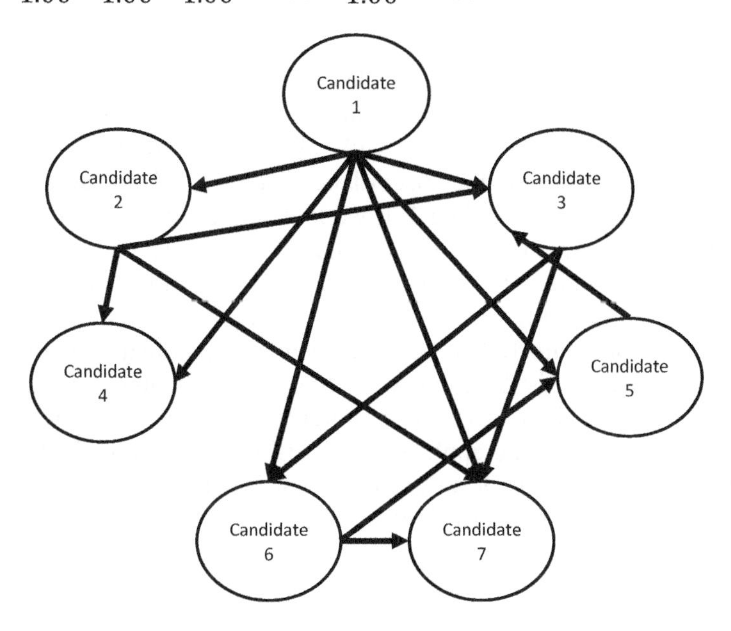

Figure 5.4. *Graph of results (Demirci and Kilic 2019)*

With the obtaining of the concordance and discordance indexes, it is possible to determine the dominance between alternatives. For this, it is necessary to set a concordance threshold and a discordance threshold. Since concordance is the "positive" information to ensure that one alternative is better than another, the concordance threshold represents the minimum amount of information required to support the hypothesis that one alternative dominates another. It is then necessary to look for the concordance indexes equal to or greater than the concordance threshold. A discordance is information that refutes this hypothesis. In this case, valid discordance indexes will be those that are equal to or less than the discordance threshold. Comparisons between alternatives for which the concordance index and the discordance index meet the defined thresholds reflect a dominance of the first alternative in the comparison over the second alternative. The concordance threshold is defined using the average of all the concordance indexes of all comparisons (all elements of the concordance matrix, rows and columns). The discordance threshold is calculated in the same way with all discordance indexes.

The graph in Figure 5.4 shows the dominances found between the seven candidates evaluated.

In the case presented, there is a clear preference for candidate 1. The application of ELECTRE shows that this candidate has the highest concordance index in comparison with all other candidates, except for candidate 4. Similarly, for the discordance indexes, all are equal to zero except for the index for candidate 4. If we look at the evaluation of candidate 1, this situation is explained by the fact that this candidate has obtained the maximum evaluation on six criteria. On the other hand, the "technical skills" criterion highlights a difference, since here candidate 4 is evaluated more favorably than candidate 1. The decision to choose candidate 1 is thus fairly obvious. The situation is different when looking at the other six candidates because in this case there is no clear preference for one candidate. In fact, the result obtained after the application of ELECTRE (Figure 5.4) does not make it possible to establish an accurate ranking for the other candidates. The analysis of the dominances shows:

– candidate 2 dominates candidates 3, 4 and 7;

– candidate 3 dominates candidates 6 and 7;

– candidate 5 dominates candidate 3;

– candidate 6 dominates candidates 5 and 7;

– candidates 4 and 7 do not show any dominance over any other candidates.

Since candidate 2 has several more dominant relationships than the other candidates, he/she can then be put at the second place in the ranking after candidate 1. However, the other candidates do not have conclusively dominant relationships, and therefore it is not possible to accurately establish the other places in the ranking. This observation reflects the phenomenon of incomparability between the alternatives, which appears to be an important limitation of the ELECTRE method.

5.3. To go further

5.3.1. *Addressing incomparability in the results*

5.3.1.1. *Toward the calculation of a comprehensive index*

To address this issue, the authors tested another computational method to define the dominances between the alternatives. For each alternative, a comprehensive concordance index and a discordance index are calculated. The comprehensive concordance index represents the sum of all outgoing concordance indexes (comparison of one alternative with respect to all others) minus the sum of all incoming concordance indexes (comparison of all alternatives with respect to an alternative). For example, the overall concordance index of the comparison of candidate 1 with candidate 2 is calculated as follows:

$$C1 = [C(1,2) + C(1,3) + C(1,4) + C(1,5) + C(1,6) + C(1,7)] - [C(2,1) + C(3,1) + C(4,1) + C(5,1) + C(6,1) + C(7,1)],$$

$$C1 = (1 + 1 + 0.85 + 1 + 1 + 1) - (0.540 + 0.236 + 0.385 + 0.529 + 0.529 + 0.529) = 5.85 - 2.748 = 3.102.$$

In this way, a concordance ranking can be defined more precisely by ordering the indexes obtained in descending order (Table 5.5).

Candidate	Comprehensive concordance index	Ranking
1	3.103	1
2	–0.119	4
3	–1.741	7
4	0.502	2
5	–0.702	5
6	0.205	3
7	–1.248	6

Table 5.5. *Ranking for the comprehensive concordance index*

The comprehensive discordance index is calculated in a similar way:

$D1 = [D(1,2) + D(1,3) + D(1,4) + D(1,5) + D(1,6) + D(1,7)] – [D(2,1) + D(3,1) + D(4,1) + D(5,1) + D(6,1) + D(7,1)]$,

$D1 = (0 + 0 + 0.358 + 0 + 0 + 0) – (1 + 1 + 1 + 1 + 1 + 1) = 0.358 – 6 = –5.642$.

Since the discordance represents the opposite information, the ranking is then given in increasing order, that is, the smaller value represents the candidate in the best position (Table 5.6).

Candidate	Comprehensive discordance index	Ranking
1	–5.641	1
2	–3.026	2
3	–0.446	4
4	–2.105	3
5	3.983	6
6	1.846	5
7	5.389	7

Table 5.6. *Ranking for the comprehensive discordance index*

Finally, to define an overall ranking, the author proposes to calculate the average placing of each candidate from the two previous rankings (Table 5.7).

Candidate	Average between the two rankings	Final ranking
1	1	1
2	3	3
3	5.5	6
4	2.5	2
5	5.5	5
6	4	4
7	6.5	7

Table 5.7. *Final ranking*

In this way, candidate 1 is shown to be the dominant candidate, but the other candidates are also given an overall ranking. In this case, the ranking of candidates is:

> Candidate 1 – Candidate 4 – Candidate 2 – Candidate 6 – Candidate 5 – Candidate 3 – Candidate 7.

This solves the limitation of incomparable results. However, the values of the concordance and discordance thresholds are not considered. That is to say, the validation and non-refutation of the outranking hypothesis are made without requiring a minimum amount of information that verifies the quality or strength of the hypothesis, to ensure that one alternative outranks another. This generates a complete ranking that allows all the alternatives to be differentiated, but this classification is less accurate due to the fact that the thresholds are not used.

One solution to this would be to change the concordance discordance thresholds so as to limit their requirements. If the goal is to obtain more dominance relationships between the alternatives, it is necessary for the thresholds to be less demanding, that is, to reduce the concordance threshold and/or increase the discordance threshold. On the other hand, if the objective is to obtain fewer dominance relationships between the alternatives, it is necessary to be more demanding with the thresholds, increasing the concordance threshold and decreasing the discordance threshold.

5.3.1.2. *Incorporating a complementary method*

In response to the same problem, Afshari et al. (2010) propose to rely on another multi-criteria analysis method to overcome this limitation of incomparability. In this case, they chose the AHP method to differentiate the two most dominant candidates. By studying the case of five candidates selected for a position at a telecommunications company, applying the ELECTRE method demonstrated that candidates 2 and 3 dominated the other three (Figure 5.5).

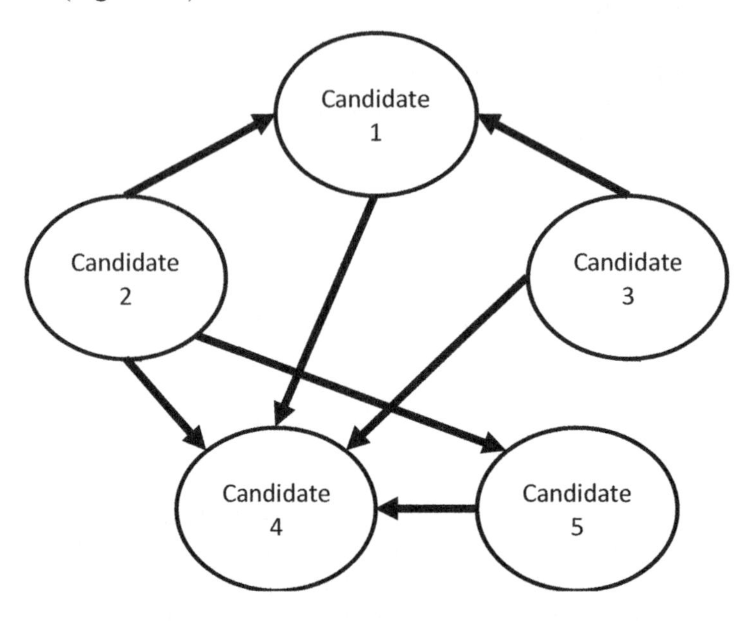

Figure 5.5. *Example of comparison of influences between criteria*

However, there is no dominant relationship between the two candidates who were the clear finalists (incomparability). For this reason, the authors applied AHP to obtain a ranking between the five alternatives and to look at the placing of candidate 2 in relation to that of candidate 3 in order to solve the incomparability that was found with ELECTRE. In this case, the classification obtained by AHP prioritizes candidate 3 over candidate 2, and thus the decision-making process results in prioritizing the hiring of candidate 3. The effectiveness of using a complementary method to differentiate between alternatives in an area of incomparability depends on how well this method is suited to the characteristics of the decision problem to be supported. For example, AHP uses the weighted average as an aggregation method, so this method is used to perform compensation. If the decision problem to be supported does not allow for any compensation to be tolerated, an application of AHP will not be recommended. In this case,

PROMETHEE (a non-compensatory method) can be considered as a complementary method.

5.3.2. *Improving the accuracy of input data: processing qualitative criteria*

In the particular context of deciding which new staff members to hire, most of the selection criteria are qualitative in nature, and the measurement of these criteria is generally not very precise. To respond to these specific concerns, Kilic et al. (2020) propose a new variation of this method, known as ELECTRE-IF, which is applied to the hiring process in a manufacturing company.

This new variation of the ELECTRE method uses the logic of a "fuzzy" evaluation. In this case study, five criteria for hiring new staff members are defined: education, experience, technical skills, personality and non-technical skills and foreign language. Three criteria are evaluated using a scale of values from 1 to 5, while the criteria "foreign language" and "experience" are evaluated by ranges of levels (Table 5.8).

Foreign language		Experience	
Evaluation	Level	Evaluation	Level
Very high	Exp > 15 years	Very high	Score > 85 points
High	10 < Exp ≤ 15 years	High	75 < score ≤ 85 points
Average	5 < Exp ≤ 10 years	Average	60 < score ≤ 75 points
Low	1 < Exp ≤ 5 years	Low	40 < score ≤ 60 points
Very low	Exp ≤ 1 year	Very low	Score ≤ 40 points

Table 5.8. *Fuzzy assessments of the foreign language and experience criteria*

Then, the evaluations obtained for the candidates for each criterion are characterized using three values: the main value, the medium range and the low value, as indicated in Table 5.9.

	Education	Experience	Technical skills	Personality	Foreign language
A	(0.25;0.75;0)	(0.75;0.2;0.05)	(1;0;0)	(0.75;0.25;0)	(0.35;0.6;0.05)
B	(0.5;0.25;0.25)	(0.9;0.1;0)	(0.25;0.5;0.25)	(1;0;0)	(0.75;0.2;0.05)
C	(1;0;0)	(0.5;0.45;0.05)	(0.25;0.5;0.25)	(0.5;0.5;0)	(0.9;0.1;0)
D	(0.5;0.25;0.25)	(0.5;0.45;0.05)	(0;1;0)	(0;0.75;0.25)	(0.75;0.2;0.05)
E	(0;1;0)	(0.35;0.6;0.05)	(0.75;0.25;0)	(0;0.75;0.25)	(0.9;0.1;0)

Table 5.9. *Primary value, medium range and low value of the criteria*

The comparison between alternatives is then carried out by studying the concordance and the discordance for the three values associated with the evaluation of each alternative. Finally, a comprehensive concordance and a comprehensive discordance are calculated to identify the dominances between the alternatives. This fuzzy logic makes it possible to incorporate imprecise or inaccurate evaluations, which are commonly found in qualitative criteria (see section 6.3).

5.4. ELECTRE: instructions for use

The ELECTRE method uses the evaluation table of the alternatives in relation to the criteria as well as the vector of the weights as its input data (Table 5.10).

	Criterion 1 (C_1)	Criterion 2 (C_2)	...	Criterion n (C_n)
Alternative 1 (A_1)	F(X_{11})	F(X_{12})		F(X_{1n})
Alternative 2 (A_2)	F(X_{21})	F(X_{22})		F(X_{2n})
...				
Alternative k (A_k)	F(X_{k1})	F(X_{k2})		F(X_{kn})
Weighting vector	W_1	W_2		W_n

Table 5.10. *Table of evaluation of alternatives in relation to the criteria (with F(X_ij) being the evaluation of alternative i in criterion j)*

To give an example of how to use the ELECTRE method, we will draw from the study by Carpitella et al. (2018) on the prioritization of critical failures in a risk analysis system. Eleven failures are evaluated using three criteria: occurrence (O), severity (S) and detection (D). Given that our goal is simply to explain the procedure for applying ELECTRE, we will work with a simplified version and consider a sample of only three failures. The evaluations of these three alternatives and the weighting vector of the criteria are presented in Table 5.11. The goal is to identify the failures that present the highest risk and to classify them according to their criticality. Therefore, occurrence and severity are criteria to be maximized and detection is, for its part, to be minimized.

	Occurrence	Severity	Detection
Failure 1	4.5	9	4
Failure 2	5.5	4	3
Failure 3	2.5	5	5
Weighting	35%	35%	30%

Table 5.11. *Example of evaluations of three failures using three criteria*

5.4.1. *Practical guide*

5.4.1.1. *Step 1: preparation of data*

Once the input data have been specified, the application of the ELECTRE method begins with the step of preparing this data. This step is divided into two parts: the normalization and the weighting of the data.

5.4.1.1.1. Normalization of data

In order to control the differences in the evaluation scales between the criteria, a normalization is carried out. In this way, the orders of magnitude used are similar. In the literature on the ELECTRE method, there are two forms of normalizations traditionally used. The first proposal is to divide each evaluation by the rank of the corresponding criterion:

$$F'(X_{ij}) = \frac{F(X_{ij})}{Max\,F(X_{ij}) - Min\,F(X_{ij})}$$

where $F(X_{ij})$ is the evaluation of alternative i for criterion j.

Considering that the columns represent the criteria, this calculation is carried out for each criterion j. The denominator within the same column j is therefore the same for all the alternatives i. This calculation method can be used for the criteria to be maximized.

If the criteria are to be minimized, the following equation is to be used:

$$F'(X_{ij}) = \frac{1/F(X_{ij})}{1/(Max\ F(X_{ij}) - Min\ F(X_{ij}))}$$

where $F(X_{ij})$ is the evaluation of alternative i for criterion j.

This formula allows us to normalize and invert the scale of the criterion simultaneously. This inversion reflects the minimization of that criterion.

With this in mind, the normalized values are presented in Table 5.12.

	Occurrence	Severity	Detection
A_1	$\frac{4.5}{5.5-2.5} = 1.5$	$\frac{9}{9-4} = 1.8$	$\frac{1/4}{1/(5-3)} = 0.5$
A_2	$\frac{5.5}{5.5-2.5} = 1.8$	$\frac{4}{9-4} = 0.8$	$\frac{1/3}{1/(5-3)} = 0.66$
A_3	$\frac{2.5}{5.5-2.5} = 0.8$	$\frac{5}{9-4} = 1$	$\frac{1/5}{1/(5-3)} = 0.4$

Table 5.12. *Normalization of assessments*

The other proposed normalization method is to divide each evaluation by the square root of the sum of all evaluations squared (as used in section 5.2).

$$F'(X_{ij}) = \frac{F(X_{ij})}{\sqrt[2]{\sum_{i=1}^{k} F(X_{ij})^2{}_{ij}}}$$

where $F(X_{ij})$ is the evaluation of alternative i for criterion j.

In the same way as for the previous normalization, the calculation is to be repeated for each criterion j. The denominator is common between all evaluations related to the same criterion j (in the same column). To work with criteria to be minimized, the concept of the inversion of scale also applies.

$$F'(X_{ij}) = \frac{1/F(X_{ij})}{1/(\sqrt[2]{\sum_{i=1}^{k} F(X_{ij})^2}_{ij})}$$

where $F(X_{ij})$ is the evaluation of alternative i for criterion j.

The main difference between the two methods of normalization is that in the second case, the order of magnitude of the normalized evaluations is between 0 and 1. The first normalization is unable to guarantee a defined interval because it depends on the original rank of the evaluations of each criterion. If the rank is low, the normalization will be between 0 and 1, and if the rank is high, the values of the normalization will be close to 1 and may possibly be higher. Thus, the second method of normalization is more effective at achieving the goal of producing more homogeneous orders of magnitude in the evaluations.

CAUTIONARY NOTE 1.– The purpose of normalization is to limit the differences in the evaluation between two criteria. The choice of the type of normalization will thus depend on the differences in scale between the criteria. Another option for normalization is the calculation proposed in the MAUT method (see Chapter 4).

5.4.1.1.2. Weighting of normalized data

The second part of the data preparation is to give weightings to the standardized assessments. The procedure is simple. Each value of the standardized table must be multiplied by the weighting for the criterion in question (specified in Table 5.11):

$$P(X_{ij}) = F'(X_{ij}) \cdot W_j$$

where:

– $P(X_{ij})$ is the normalized and weighted evaluation of alternative i for criterion j;

– $F'(X_{ij})$ is the standardized evaluation of alternative i for criterion j;

– W_j is the weighting defined for criterion j.

The results from the data preparation step are then used to create the normalized and weighted table of evaluations (Table 5.13).

	Occurrence	Severity	Detection
A_1	$1.5 \times 0.35 = 0.525$	$1.8 \times 0.35 = 0.63$	$0.5 \times 0.3 = 0.15$
A_2	$1.8 \times 0.35 = 0.63$	$0.8 \times 0.35 = 0.28$	$0.66 \times 0.3 = 0.198$
A_3	$0.8 \times 0.35 = 0.28$	$1 \times 0.35 = 0.35$	$0.4 \times 0.3 = 0.12$

Table 5.13. *Normalized and weighted values*

5.4.1.2. *Step 2: calculation of concordances*

The concordance and discordance indexes depend directly on the comparison between alternatives. Both for the concordance and the discordance, a square matrix must be constructed, where the size of the matrix is determined by the number of alternatives. The main diagonal of these two matrices will be empty, since the comparison of an alternative with itself is not considered.

In the concordance matrix, each element represents the concordance index of a comparison between two alternatives. The matrix is read in the following way: the concordance index of the alternative of the line with respect to the alternative of the column.

Concordance represents the information to ensure that one alternative is actually better than another. In the comparison between two alternatives, it is therefore necessary to identify the criteria for which the first alternative in the comparison is better than the second. The criteria that meet this condition will add their weightings to the concordance index from this comparison. The criteria that do the opposite (where the second alternative of the comparison is better than the first) do not contribute anything to the concordance index. Finally, the criteria that show equal values between the two alternatives will contribute half of their weight to the corresponding concordance index. Some authors propose that, when the values are equal, the full weight of the criterion should be added. However, this variation in the treatment of equally valued assessments has little or no impact on the results. In this case, we will work with an addition of 50% of the weight of the criterion in the case of a tie.

The concordance index is therefore defined by the following equation:

$$C\ (i; i+1) = \sum_{j=1}^{n} W_j\ si\ P(X_{ij}) > P(X_{i+1j}) + \sum_{j=1}^{n} 0.5W_j\ if\ P(X_{ij}) = P(X_{i+1j}),$$

where:

– $P(X_{ij})$ is the normalized and weighted evaluation of alternative i for criterion j;

– W_j is the weighting defined for criterion j.

The rule to use half the weight in the case of equal values makes it possible to establish a relationship between the cross-evaluated concordance indexes because their sum must be equal to 1. That is, if the concordance index C(i; i + 1) considers the sum of the weights of the criteria for which alternative i is better than alternative i + 1, plus half of the weightings of the criteria for which the alternatives are equal; then by contrast, the opposite concordance index C(i + 1; i) will consider the sum of the weights of the criteria for which the alternative i + 1 is better than the alternative i, plus half of the weightings of the criteria for which the alternatives are equal. Thus, the total values of the weightings of all the criteria will be represented by these two concordance indexes. The sum of index C(i ; i + 1) + index C(i + 1; i) will thus be equal to 1.

The concordance matrix of our case study looks like this:

$$\begin{pmatrix} - & 0.35 & 1 \\ 0.65 & - & 0.65 \\ 0 & 0.35 & - \end{pmatrix}.$$

CAUTIONARY NOTE 2.– The calculation of the concordance indexes does not depend on data normalization; it is therefore possible to invert the order of the first two steps.

5.4.1.3. *Step 3: calculation of discordances*

In the same way as for the concordance matrix, the discordances are calculated using comparisons between the alternatives. Therefore, the discordance matrix will also be square, and its size will depend on the number of alternatives considered. While concordance refers to the information used to confirm the hypothesis where one alternative is deemed better than another, discordance considers the information that makes it possible to refute this hypothesis. This time, the discordance index seeks to quantify the intensity of the information needed to refute the hypothesis. The calculation for this index is based on the differences in the evaluation between the two alternatives compared according to each criterion. First, it is necessary to identify the differences for which the first alternative is equal to or worse than the second alternative of the comparison, since these differences reflect discordances. Then, these discordances must be compared with respect to the global differences of these alternatives on all the criteria. Therefore, the discordance index is defined as the ratio of the maximum difference of the criteria for which the first alternative in the comparison is equal to or less favorable than the second, divided by the maximum difference between these alternatives without taking into account the dominance of one over the other. Thus, our goal is to find the maximum difference between the two alternatives, regardless of whether the first alternative is more desirable, less desirable or equal to the second:

$$\mathrm{D}(i;\, i{+}1) = \frac{\max |\, (P(X_{ij}) - P(X_{i+1j})| (\text{où } P(X_{ij}) \leq P(X_{i+1j}))}{\max |\, P(X_{ij}) - P(X_{i+1j})|}$$

where $P(X_{ij})$ is the normalized and weighted evaluation of alternative i for criterion j;

The discordance matrix is therefore:

$$\begin{pmatrix} - & 0.3 & 0 \\ 1 & - & 0.2 \\ 1 & 1 & - \end{pmatrix}.$$

In this case, the only relationship between the opposite discordance indexes is that they share a common denominator and that at least one index for both will be equal to 1 (in fact, in one of the cases, the numerator will be equal to the denominator).

CAUTIONARY NOTE 3.– The opposite discordance indexes share the same denominator and at least one of the two will be equal to 1, except in the case where the alternatives are equal on all criteria. In the discordance matrix, at least half of the elements are equal to 1.

5.4.1.4. *Step 4: definition of concordance and discordance thresholds*

As the objective of the method is to collect information to validate the outranking hypothesis of an alternative compared to the others. The definition of a concordance threshold is the minimum amount of information required to support the hypothesis. A dominant concordance matrix is constructed by setting the concordance indexes that meet this threshold as equal to one. For example, if the threshold is set from the average of the set of concordance indexes (0.5), the matrix will be:

$$\begin{pmatrix} - & 0 & 1 \\ 1 & - & 1 \\ 0 & 0 & - \end{pmatrix}.$$

On the other hand, the discordance threshold represents the maximum amount of information allowed that would not refute the hypothesis. In this case, the valid discordance indices will be the indexes equal to or less than the threshold, and will be represented by 1 in the dominant discordance matrix:

$$\begin{pmatrix} - & 1 & 1 \\ 0 & - & 1 \\ 0 & 0 & - \end{pmatrix}.$$

In this step, it is necessary to identify the comparisons of alternatives where the concordance index is equal to or greater than the concordance threshold and where the discordance index is equal to or less than the discordance threshold. If a comparison meets these two conditions, it can be concluded that the first alternative of the comparison dominates the second. If none or only one of the two conditions are met, it is not possible to establish a dominance between the alternatives. The confirmed dominances are represented by 1 in the last matrix, named the aggregate dominance matrix:

$$\begin{pmatrix} - & 0 & 1 \\ 0 & - & 1 \\ 0 & 0 & - \end{pmatrix}.$$

The setting of the concordance and discordance thresholds can be done by the decision maker or by calculating the average of all the corresponding elements of the matrix. In the example given in this section, the concordance threshold concordance is set to 0.5 because the average of all terms in the concordance matrix is equal to 0.5 (the main diagonal is not taken into account for the calculation of the average):

$$\text{concordance threshold} = [0.35 + 1 + 0.65 + 0.65 + 0 + 0.35]/6 = 0.5.$$

CAUTIONARY NOTE 4.– The most common thresholds are 0.5 for the concordance threshold, and between 0.7 and 0.8 for the discordance threshold.

CAUTIONARY NOTE 5.– If we carry out a sensitivity analysis in the aggregate dominance matrix, by seeking to find more dominance relationships, the thresholds must be made less demanding by decreasing the concordance threshold and by increasing the discordance threshold. On the other hand, if the objective of the sensitivity analysis is to limit the dominance relationships, the thresholds must be made more demanding by increasing the concordance threshold and decreasing the discordance threshold.

5.4.1.5. *Step 5: construction of the ELECTRE graph*

The graph of ELECTRE represents the dominances between alternatives identified in the previous step. All 1s in the aggregated dominance matrix indicate a dominance of the alternative given in the row with respect to the alternative given in the column. On the graphic, this dominance will be represented by an arrow.

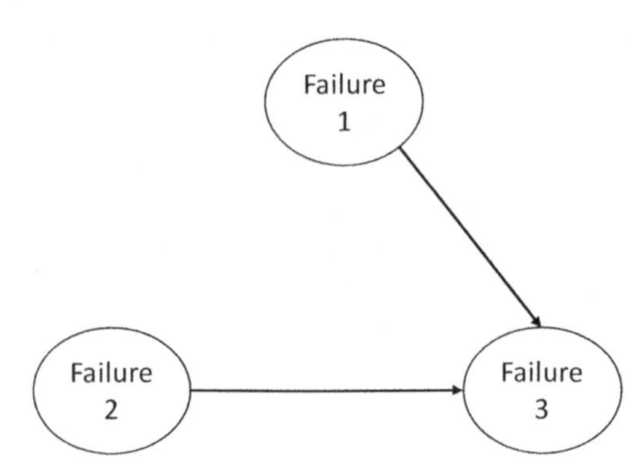

Figure 5.6. *Graph of the results of the example*

In this graph, the arrows indicate the dominance of the alternative at the origin of the arrow versus the alternative at its tip. For example, in Figure 5.6, alternative 1 dominates alternative 3, and alternative 2 dominates alternative 3. There are no dominance relationships between alternatives 1 and 2.

CAUTIONARY NOTE 6.– The graph can be interpreted by reading the direct dominance relationships, but also using the principle of transitivity: for example, if alternative A dominates alternative B, and alternative B dominates alternative C; we can deduce using the transitive property that alternative A dominates alternative C.

5.4.2. *Illustration of related free software*

As with the other multi-criteria analysis methods, various software programs have been developed to facilitate the application of ELECTRE. Most of them are commercial software, but in this book, we will give an example using the freeware program Decision Radar, which offers applications of different multi-criteria analysis methods, including ELECTRE.

The application of ELECTRE using Decision Radar allows to define a problem with a maximum of seven alternatives and seven criteria. Now, we will return to the previous example and define the alternatives (in our case, the three failures) (Figure 5.7).

Set Choices

Enter your choices up to 7

Choices	Name
1	Failure 1
2	Failure 2
3	Failure 3

Figure 5.7. *Entering alternatives into the Decision Radar software*

Next, the decision criteria should be defined: the name of the criterion, the direction (for those to be maximized without checking the "negative" box/for those to be minimized by checking the "negative" box) and the weighting of the criterion and its nature (qualitative or quantitative) (Figure 5.8).

Set Indicators

Enter your indicators up to 7

Choices	Name	Negative	Weight	Qualitative
1	Occurrence	☐	35	☐
2	Severity	☐	35	☐
3	Detection	☑	30	☐

Figure 5.8. *Entering criteria into the Decision Radar software*

Finally, we must enter the evaluations of each alternative in relation to each criterion as shown in Figure 5.9.

Set Decision Matrix

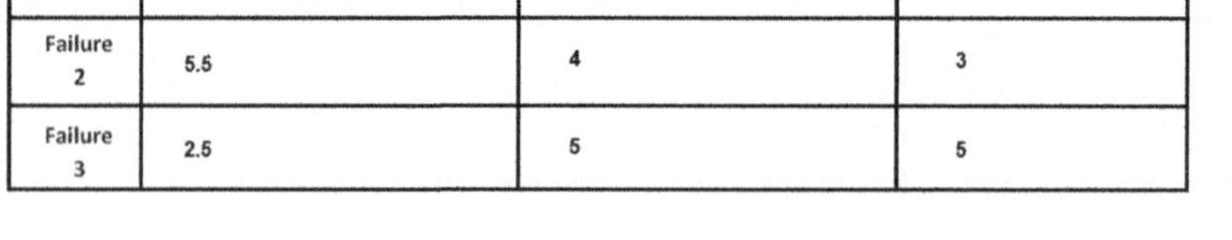

	Occurrence	Severity	Detection
Failure 1	4.5	9	4
Failure 2	5.5	4	3
Failure 3	2.5	5	5

Figure 5.9. *Entering evaluations into the Decision Radar software*

After clicking on the "calculate with ELECTRE Method" button, the software will solve the problem by applying ELECTRE. The primary result of the software is the generating of dominance relationships between the alternatives. In our example, this is the dominance of alternatives 1 and 2 over alternative 3 (Figure 5.10).

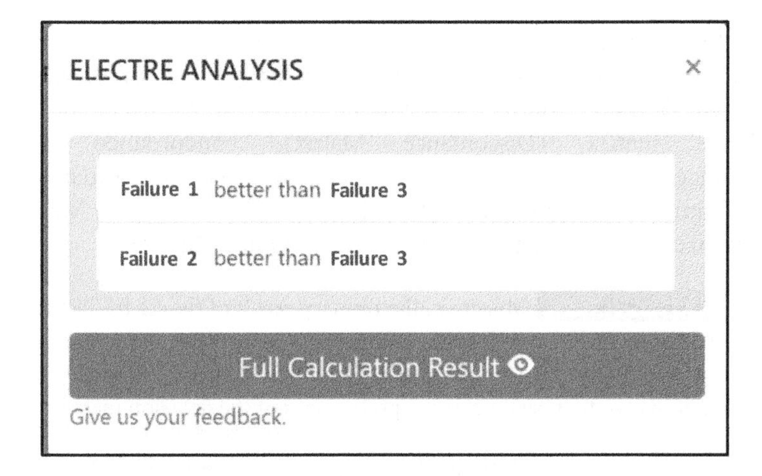

Figure 5.10. *Results given by the Decision Radar software*

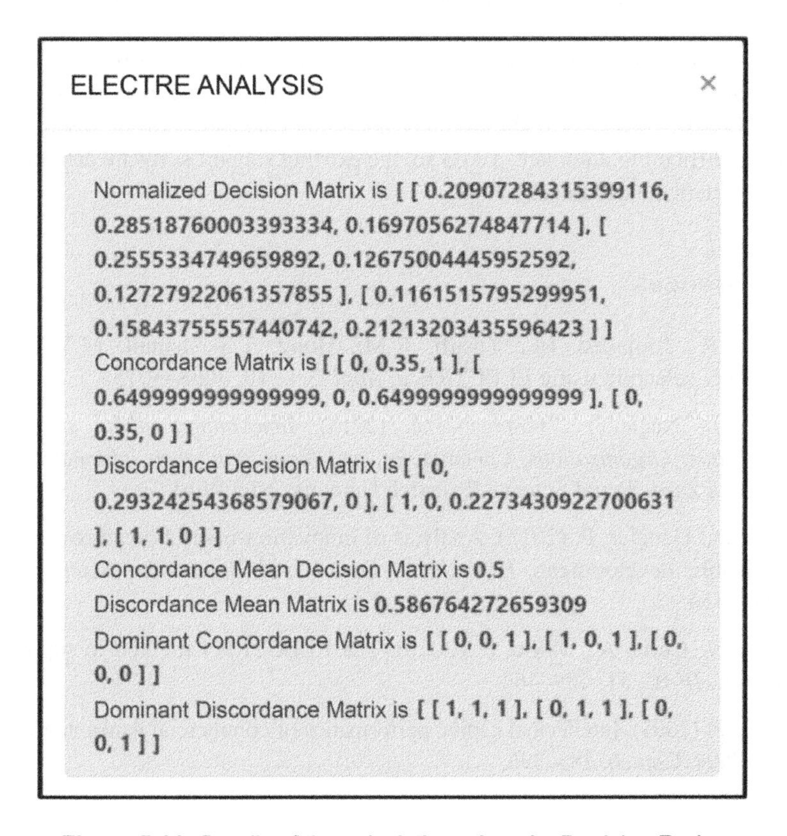

Figure 5.11. *Details of the calculations done by Decision Radar*

The "Full Calculation Result" option allows you to view the most important calculations of the method: matrix of normalized values (Normalized Decision Matrix), concordance matrix (Concordance Matrix), discordance matrix (Discordance Matrix), concordance threshold (Concordance Mean Matrix), discordance threshold (Discordance Mean Matrix), dominant concordance matrix (Dominant Concordance Matrix) and dominant discordance matrix (Dominant Discordance Matrix).

ELECTRE AT A GLANCE.– Objective: the primary result of the ELECTRE method is the establishment of dominance relationships between alternatives. These relationships are represented graphically using the dominance graph. Dominance shows the alternative that best meets the preferences of the decision maker.

Unique features: the ELECTRE method works on the principle of outranking, based on pairwise comparisons between alternatives. This comparison is carried out criterion by criterion, allowing for compensation phenomenon to be managed.

Limits: the most important limit is that in some cases it is not possible to establish a ranking; however, this limit has been addressed by the new versions of ELECTRE (II, III and IV). On the other hand, the calculation of the discordance matrix is difficult to automate. To do so, the existing support software can provide very important contributions.

5.5. References

Afshari, A.R., Mojahed, M., Yusuff, R.M., Hong, T.S., Ismail, M.Y. (2010). Personnel selection using ELECTRE. *J. Appl. Sci.*, 10, 3068–3075.

Altman, E.J., Nagle, F., Tushman, M. (2014). *Innovating without Information Constraints: Organizations, Communities, and Innovation when Information Costs Approach Zero*. Social Science Research Network, New York.

Bircan, I. and Gençler, F. (2015). Analysis of innovation-based human resources for sustainable development. *Procedia – Social and Behavioral Sciences*, 195, 1348–1354.

Carpitella, S., Certa, A., Mario, E. (2018). The ELECTRE I method to support the FMECA. *IFAC*, 51, 459–464.

Chen Goh, P. (2005). Intellectual capital performance of commercial banks in Malaysia. *J. Intellect. Cap.*, 6, 385–396.

Chiu, H.H. (2018). Employees' intrinsic and extrinsic motivations in innovation implementation: The moderation role of managers' persuasive and assertive strategies. *J. Chang. Manag.*, 18, 218–239.

Demirci, E. and Kilic, H.S. (2019). Personnel selection based on integrated multi-criteria decision making techniques. *Int. J. Adv. in Eng. Pure Sci.*, 2, 163–178.

Dursun, M. and Karsak, E.E. (2010). A fuzzy MCDM approach for personnel selection. *Expert Syst. Appl.*, 37, 4324–4330.

Govindan, K. and Jepsen, M.B. (2016). ELECTRE: A comprehensive literature review on methodologies and applications. *Eur. J. Oper. Res.*, 250, 1–29.

Güngör, Z., Serhadlıoğlu, G., Kesen, S.E. (2009). A fuzzy AHP approach to personnel selection problem. *Appl. Soft Comput.*, 9, 641–646.

Ijadi Maghsoodi, A., Riahi, D., Herrera-Viedma, E., Zavadskas, E.K. (2020). An integrated parallel big data decision support tool using the W-CLUS-MCDA: A multi-scenario personnel assessment. *Knowl.-Based Syst.*, 195, 105749.

Kilic, H.S., Demirci, A.E., Delen, D. (2020). An integrated decision analysis methodology based on IF-Dematel and IF-ELECTRE for personnel selection. *Decis. Support Syst.*, 137, 113360.

Lenihan, H., McGuirk, H., Murphy, K.R. (2019). Driving innovation: Public policy and human capital. *Res. Policy*, 48, 103791.

Liu, M., Li, M., Zhang, T. (2012). Empirical research on China's SMEs technology innovation engineering strategy. *Syst. Eng. Procedia*, 5, 372–378.

Madrid, H.P., Diaz, M.T., Leka, S., Leiva, P.I., Barros, E. (2018). A finer grained approach to psychological capital and work performance. *J. Bus. Psychol.*, 33, 461–477.

Maystre, L.Y., Pictet, J., Simos, J. (1994). *Méthodes multicritères ELECTRE : description, conseils pratiques et cas d'application à la gestion environnementale.* Presses polytechniques et universitaires romandes, Lausanne.

Roy, B. (1968). Classement et choix en présence de points de vue multiples. *RAIRO (Recherche Opérationnelle)*, 2, 57–75.

Roy, B. (1976). A conceptual framework for a normative theory of "decision-aid". Report, Centre national de l'entrepreneuriat, Laboratoire de management sientifique et aide à la décision.

Sang, X., Liu, X., Qin, J. (2015). An analytical solution to fuzzy TOPSIS and its application in personnel selection for knowledge-intensive enterprise. *Appl. Soft Comput.*, 30, 190–204.

6

Knowledge Management in the Supply Chain: An Application of TOPSIS

Knowledge is recognized as a key factor for the success of companies. At the core of innovation processes, knowledge offers a particularly strong added value for an organization by allowing it to respond to and/or anticipate rapid changes in society and to register as a key player within the global competition.

However, in order for knowledge to become a real success factor and a driver of innovation, it is essential to put in place an approach for managing knowledge that is structured and based simultaneously on the search for storage and sharing of information. It is precisely this approach that makes it possible to capitalize on and transform data and facts into knowledge, and therefore into added value. By implementing specific knowledge management practices, an organization can implement the strategic exploitation of available knowledge to create a real competitive advantage. These practices can then impact all the processes of the organization, as well as the entire supply chain.

Thus, the real challenge of knowledge management lies in the ability of organizations to put in place concrete and sustainable actions for which it is generally necessary to remove the barriers to adopting them over the long term.

To illustrate this issue, we will draw on a study carried out by Patil and Kant (2014) aimed at prioritizing the operational solutions to be put in place to remove barriers to the adoption of knowledge management practices within the supply chain of a company that manufactures hydraulic valves and pumps. This article proposes an application of the TOPSIS method (technique for order performance by similarity to ideal solution) developed by Hwang and Yoon in the 1980s.

6.1. Context and challenges in decision-making

Knowledge is a strategically decisive resource for companies (Talbi 2018). It plays a crucial role in the innovation process by conserving the experiences from previous projects. Innovation is a cyclical process that repeats itself with each project, and questioning the successes and failures of these projects helps to improve performance. This means drawing from existing knowledge and its evolution over time in order to better approach further evolutions of this knowledge in the future (Benhamou et al. 2002). This phenomenon is known as knowledge management. It consists of organizing the collection, preservation, valuation and creation of strategic knowledge for the company. It cannot be reduced solely to the transmission of information. To develop a dynamic of innovation, it is necessary to capitalize on the knowledge acquired, on the one hand, to reveal the conditions that have been the factors that trigger innovation and, on the other hand, to open the way for new knowledge by initiating a process of learning and continuous improvement. This capitalizing consists of ensuring that the knowledge and experience acquired during projects will remain in the company's possession, regardless of any change in people or management, and that it can be reused for future projects.

In this section, we will specifically illustrate the importance of knowledge management in the innovation process through the particular case of supply chain management.

6.1.1. *Knowledge management in the supply chain*

The concept of the supply chain, as considered in this chapter, refers to a successive set of value-added activities that allow for raw materials to be transformed into a finished product to be used by customers (Gereffi and Fernandez-Stark 2018). Thus, there are potentially many players involved in the supply chain, who are geographically dispersed and both internal and external (suppliers, distributors, subcontractors, customers, etc.) to the organization in question.

Knowledge management in the supply chain, and more particularly the transfer of this knowledge to all the actors concerned, is a significant challenge for companies. Indeed, the notion of performance as it relates to the knowledge management is difficult to evaluate compared to other key performance indicators (KPIs) used in logistics. Thus, supply chain actors generally have little access to information, but despite this, these same actors consider knowledge management a secondary issue. This explains, in particular, the low participation of supply chain actors in the knowledge management approach, the low amount of resources dedicated to it, as well as the barriers to the adoption of good practices that would improve it. However, promoting the transfer of knowledge between supply chain actors can be accompanied by many benefits. By creating a more open and collaborative working environment, the supply chain and its actors gain adaptability and agility, and this translates into an increase in competitiveness. In the same way, better knowledge management leads to an optimized use of resources and better quality control of products. Of course, this is a potential source of innovation.

In the era of digitalization, one of the emblematic illustrations of this is the concept of "Industry 4.0". Industry 4.0 can be defined as a kind of spinal cord, integrating human factors, technological tools, lines production and processes across the organizational boundaries of a company, forming a new supply chain that is smart, connected and agile (Schumacher et al. 2016). Sensors, Big Data, robotics, data analysis and control are all concrete technological solutions put in place to extract, interpret and disseminate knowledge within the supply chain.

However, whether in a digitalization approach or for any other initiative related to knowledge management, its benefits can only be truly perceived if the organization overcomes the difficulty of moving knowledge management from an operational and secondary dimension to a strategic and properly valued one. While the top management plays a crucial role in creating this favorable environment for the adoption of knowledge management practices, it is nevertheless a real problem of the management of changes for which the actors of the supply chain must be supported. Thus, one of the key challenges of knowledge management in the supply chain is to remove existing barriers to the adoption of new practices, progressively and in a concrete way. We will illustrate this problem through an application case in the Indian manufacturing industry for which a decision model has been constructed using the TOPSIS methodology.

6.1.2. *Definition of the decision model*

The case we will study here is inspired by the article by Patil and Kant (2014) in which they present a case study within an Indian company that designs and manufactures hydraulic pumps and valves for industrial use. This company has 150 employees and collaborates with more than 26 suppliers and distributors. The goal of this company is to turn its knowledge into a competitive advantage through the adoption of knowledge management practices in its supply chain. In carrying this out, it faces barriers related to the adoption of these new practices by the actors of its supply chain. Thus, it is seeking, on the one hand, to identify concrete solutions to overcome these barriers, and on the other hand, to prioritize them in such a way as to facilitate the step-by-step adoption of these solutions, avoiding the simultaneous mobilization of resources.

This decision model was built in collaboration with different groups of experts and combined with an in-depth study of the academic literature. An initial working group made up of 15 decision makers within the company was put in place. This group included senior managers, IT managers, members of the supply chain and customers. Based on an initial theoretical selection of barriers to the adoption of knowledge management practices within the supply chain, they then resulted in the creation of five main categories:

– *strategic barriers* (SBs) refer to the difficulties associated with the promotion of knowledge management in the supply chain and a good understanding of its potential strategic advantages: lack of a plan for the integration of risk management into the company's processes, difficulty in evaluating its performance, a weak definition of the roles of the actors involved, a lack of dedicated resources, a lack of involvement by management and overall vision, etc.;

– *organizational barriers* (OBs) refer to the lack of an organizational structure to create and share knowledge: a lack of communication, an exclusively top-down information flow, information retention and a lack of spaces dedicated to reflection and exchange;

– *technological barriers* (TBs) refer to the lack of technical infrastructures supporting the creation, storage and dissemination of information: centralization of data, difficulty in capitalizing tacit knowledge, a lack of security in the dissemination of information, a lack of technical assistance for all members of the supply chain, etc.;

– *cultural barriers* (CBs) refer to the dynamics of collaboration within the company: a lack of trust and commitment, managerial mentality, hierarchical corporate culture, lack of recognition and motivation, multicultural environment, sharing spirit not present in teams, the opportunistic behavior of members of the supply chain, etc.;

– *individual barriers* (IBs) refer to the behaviors and reticence of employees regarding knowledge management: the fear of sharing false information, a lack of available time, attitudes toward learning, a feeling of ownership of information, poor communication skills, lack of training, etc.

Then, a second group of experts was called on to identify and select concrete solutions to remove the barriers highlighted above. This time, this second working group was made up of five experts from outside the company, specialists in the supply chain and knowledge management. Ten solutions were selected and described (Table 6.1).

In this way, the decision model built for this application includes four decision criteria (barriers to the adoption of knowledge management practices in the supply chain) and 10 alternatives to prioritize (the concrete solutions to be implemented in order to remove these barriers). The TOPSIS method was chosen to answer these decision-related issues. The objective was to determine which concrete solutions have the greatest capability to overcome the identified barriers and then facilitate the adoption of knowledge management practices within the company and its supply chain.

Selected solutions	Description
S1 – Establish positive leadership	Motivate employees to adopt knowledge management practices by offering financial and technical support for the actions carried out, by establishing regular checks and by establishing long-term planning. Post-adoption audits can also be conducted. These actions reflect an interest, involvement and active monitoring by the management.
S2 – Encourage teamwork and transparency	Build a safe and trustworthy collective working environment to exchange and improve knowledge within the supply chain. Promote exchanges by adopting an "open door" policy promoting transparency. This eliminates the difficulties of information flow from one level to another and ensures agility, adaptability and the alignment of the supply chain.
S3 – Strengthen cultural cohesion and cooperation	This means creating a supply chain where employees at all levels support the company's core values and understand their contribution as an individual and as a member of the team they are part of. This involves creating a set of concepts and principles that help determine which behaviors help or hinder the improvement of the supply chain's performance.
S4 – Put incentives in place	One effective way to encourage employees to share their knowledge within the supply chain is to recognize when they do their work well, through adequate incentives and rewards. Incentive and reward systems are a formal system used to promote or encourage specific actions or behaviors of a group of people for a defined period of time.
S5 – Design an outsourcing strategy	Outsourcing an internal business process to a third-party organization (outsourcing) encourages the supply chain to become a "systems integrator", which manages and coordinates a network composed of the best suppliers. This makes it possible to improve the integration of knowledge within the supply chain.
S6 – Create mutual learning	Supply chain decision makers and experts exchange regularly on issues of common interest to carry out effective knowledge sharing. This mutual learning process makes it possible to improve coordination within the supply chain, as well as within the decision-making process.
S7 – Use digital tools	For both the dissemination of knowledge and knowledge storage or for remote collaborative work, the ability to use information technology and Internet tools is a valuable asset. These make it possible to store, retrieve, transmit and change data across organizational boundaries. There are many applications that promote the flow of information within the supply chain, including enterprise resource planning (ERP) systems, radio frequency identification (RFID) devices, multi-agent systems, the semantic web, video conferencing, etc.

Selected solutions	Description
S8 – Build partnerships	Collaborative projects with suppliers to foster continuous improvements, strategic alliances with members of the supply chain to pursue an agreed-on set of objectives or to meet an essential business need, or even to create a virtual company with another entity to share skills and/or resources, are all examples of partnerships that can improve knowledge management within the supply chain.
S9 – Strengthen the data collection process	Intellectual property and customer relationship management (CRM) systems are valuable tools for accelerating learning and deriving benefits from knowledge within the supply chain. Intellectual capital represents the sum of all the knowledge that the supply chain makes use of in order to obtain a competitive advantage. CRM allows for the collection of data to improve the relationship between the supply chain and its customers. Finally, the acquisition of knowledge and feedback from supply chain workers also makes it possible to formulate and adapt its configuration to meet operational specifications.
S10 – Disseminate collaborative practices	Collaborative practices place greater focus on exchanges of information and knowledge sharing as key success factors for the supply chain. Vendor managed inventory (VMI), effective consumer response (ECR), enhanced web reporting (EWR) or collaborative planning and forecasting and replenishment (CPFR) are all examples of how to gradually build knowledge within the supply chain.

Table 6.1. *Selected solutions*

6.2. The TOPSIS method

The TOPSIS method was developed by Hwang and Yoon in the 1980s. It is based on the principle that the alternative chosen by the decision maker must have the shortest relative geometric distance to the ideal solution (the one that is the best in terms of all the criteria considered) and the longest geometric distance relative to the worst solution (the one that scores the worst on all the criteria).

Thus, the different steps of the TOPSIS method are intended to achieve a compromise between minimizing the distance to the ideal situation and maximizing the distance to the unwanted situation (Figure 6.1). This compromise results in the calculation of a proximity index, which thus makes it possible to rank the alternatives from the most favorable to the most unfavorable.

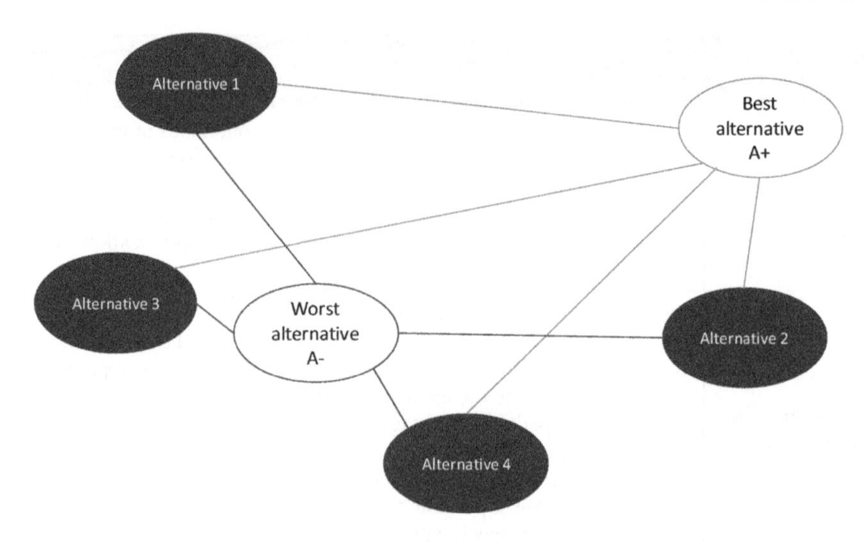

Figure 6.1. *Illustration of distances to the best (A+) and to the worst alternative (A-)*

Thus, TOPSIS is a multi-criteria analysis ranking method, as it produces a ranking of the alternatives based on the calculation of an aggregated score from the criteria and preferences defined by the decision maker. The main advantages of this method are its ease of implementation and its intuitive nature. On the other hand, it is important to be aware that it is also partially compensatory in nature, which is a product of the computational steps on which it is based (i.e. its use of weighted sums).

Moreover, one limitation of this method is that in cases where all the alternatives considered are bad in view of the preferences of the decision maker, the method then proposes the best possible action among bad options, which may prove unsatisfactory. Despite these issues, TOPSIS remains one of the most widely used decision support methods, thanks in particular to its computational advantages.

6.2.1. *Methodological concepts*

6.2.1.1. *The best and worst alternatives*

The best and worst alternatives are fictitious alternatives representing an ideal value to be achieved and an undesirable value to be avoided. They are traditionally rated as A+ and A-. Therefore, they are not alternatives that are integrated into the decision problem. These are additional alternatives which act as theoretical frames of reference for introducing the notion of distance, which is central to the application of the TOPSIS method.

It is thus necessary to first define which of the alternatives is the best and which is the worst. This stage directly depends on the definition of the criteria of the decision problem and the goal associated with them.

These objectives can often vary widely (Figure 6.2):

– maximizing a criterion: the decision maker systematically prefers a solution that offers an evaluation with the highest mathematical value (e.g. a criterion for lifespan);

– minimizing a criterion: the decision maker prefers a solution where the evaluation represents the lowest mathematical value (e.g. a criterion for cost);

– the definition of a target value: the decision maker prefers an evaluation whose value is closest to an intermediate value deemed to be ideal (e.g. a criterion for temperature or size). The solutions whose evaluations are closest to this ideal will then be preferred on this criterion, while the values furthest from the target, both lower and higher, will potentially be deemed undesirable.

Thus, the definition of the best and worst alternatives for the application of TOPSIS directly depends on the goal associated with each criterion that is part of the decision problem. If all of the criteria are to be maximized (i.e. we systematically prefer the value where the evaluation is the highest), then the best alternative will be represented by the highest evaluation scores on all of the criteria, while the worst solution will be represented by the lowest evaluation scores for all of the criteria.

On the other hand, in the case where the criteria set includes at the same time criteria to be maximized, to be minimized, and target values, then the best approach is to choose the value that represents either the best or the worst alternative on a case-by-case basis for each criterion and then combine all of them. For example, for an ordinary decision maker, the best alternative in the case of buying a home would be a property that is as cheap as possible, has the highest energy performance on the market, while being as close as possible to a surface area of 70 m^2. This house probably does not exist, but it represents an ideal to be achieved, from which the alternatives considered must be the closest as possible.

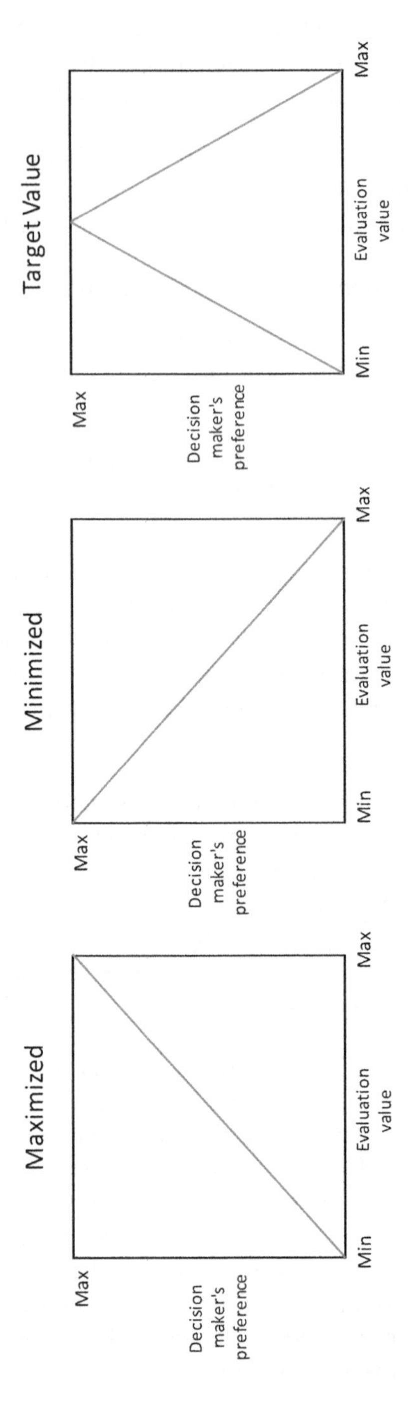

Figure 6.2. *The objectives associated with the criteria for the decision*

6.2.1.2. *The notion of distance and proximity*

As mentioned earlier, the TOPSIS method is based on a distance calculation allowing for a solution to be chosen that is as close as possible to the ideal alternative (A+) and that is as far away as possible from the worst solution (A-). Traditionally, the method is based on a Euclidean distance calculation. On the one hand, it is a question of calculating the distance separating each solution from the best alternative A+ according to the following formula:

$$D_j^{+} = \sqrt{\sum_{i=1}^{n} [E_{norm,i}(\mathrm{A}+) - E_{norm,i,j}]^2}$$

where:

– j is the alternative considered;

– i is the criterion considered;

– n is the number of criteria in the decision problem;

– $E_{norm,i}(\mathrm{A}+)$ is the evaluation value of alternative A+ for criterion i;

– $E_{norm,i,j}$ is the evaluation value of alternative j for criterion i;

Similarly, the Euclidean distance separating each solution from the worst alternative A- is calculated using the following formula:

$$D_j^{-} = \sqrt{\sum_{i=1}^{n} [E_{norm,i}(\mathrm{A}-) - E_{norm,i,j}]^2}$$

where:

– j is the alternative considered;

– i is the decision criterion considered;

– n is the number of criteria in the decision problem;

– $E_{norm,i}(\mathrm{A}-)$ is the normalized evaluation value of alternative A- for criterion i;

– $E_{norm,i,j}$ is the normalized evaluation value of alternative j for criterion i.

Finally, from the calculated distances, it is possible to construct a closeness coefficient CC that allows for highlighting the solutions that are both the closest to the best alternative A+ and also the furthest from the worst alternative A-. This coefficient is calculated as follows:

$$CC_j = \frac{D_j^-}{D_j^- + D_j^+}$$

where:

$-j$ is the alternative considered;

$-D_j^-$ is the Euclidean distance separating solution j from the worst alternative A-;

$-D_j^+$ is the Euclidean distance separating solution j from the best alternative A+.

This coefficient is the highest (approaching 1) when the distance D_j^+ is low (the solution is therefore close to the best alternative) and the distance D_j^- is large (the solution is therefore farther from the worst alternative). Similarly, this proximity coefficient is the lowest (approaching 0) when the distance D_j^- is small, and the distance D_j^+ is large.

6.2.2. *Application of the method*

In order to clarify these methodological notions, we will now apply the different steps of the TOPSIS method to the case study that we have chosen to address: *how to limit the barriers to the adoption of knowledge management practices in the supply chain for a company X.* For this, we drew from the research work of Patil and Kant (2014). We will therefore detail the characteristic steps of the TOPSIS method presented in Figure 6.3.

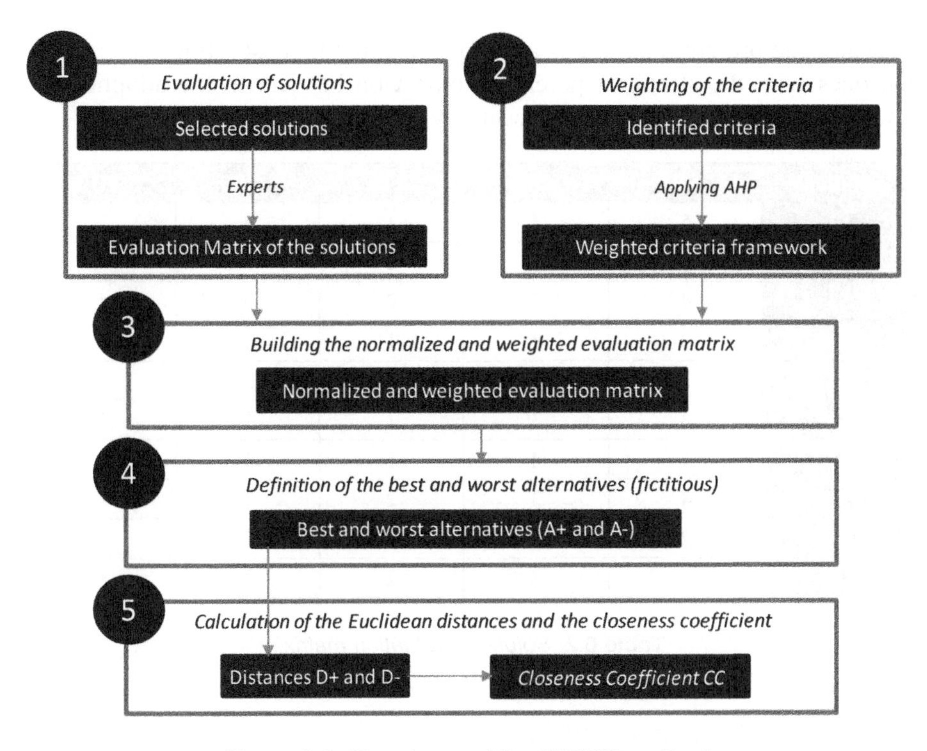

Figure 6.3. *The stages of the TOPSIS method*

6.2.2.1. *Evaluation of solutions*

Potential solutions identified to facilitate knowledge management within the supply chain of company X were evaluated by a group of five experts made up of supply chain and knowledge management specialists. Each of the 10 selected solutions was evaluated according to the five categories used in our decision model: strategic, organizational, technological, cultural and individual barriers. The experts therefore agreed to assign five values to each solution, representing its ability to overcome each of the identified barriers:

1 = very low ability to address the barrier;

2 = low ability to address the barrier;

3 = moderate ability to address the barrier;

4 = high ability to address the barrier;

5 = very high ability to address the barrier.

Thus, the preferred solutions are those with the highest ratings. These are the ones with the strongest potential impact on barriers to the adoption of knowledge management within the supply chain (Table 6.2).

Solutions	SB	OB	TB	CB	IB
S1	5	4	1	3	2
S2	4	5	2	5	4
S3	4	2	2	5	4
S4	3	3	1	5	5
S5	3	3	3	2	2
S6	3	3	1	3	3
S7	1	4	5	1	1
S8	4	3	2	1	1
S9	2	3	4	1	1
S10	1	4	2	3	3

Table 6.2. *Solution evaluation matrix*

6.2.2.2. *Weighting of the criteria*

The step of weighting the criteria is a central step in the modeling of decision makers' preferences and reflects the relative importance of the selected criteria (and thus their impact on the decision). In our case, the AHP method was used (see Chapter 2). For this, a group of 15 decision makers was consulted. Each of the decision makers was asked about the importance they gave to each barrier in the decision model.

The essential elements of the AHP method consisted of questioning decision makers using a pairwise comparison, asking them to indicate which of the two barriers within a pair they consider as more important. The Saaty scale (1980) was then used to quantify the importance. For a given pair of criteria, the decision makers would then take a position on one side or the other of the Saaty scale by qualifying the degree of importance of one barrier relative to the other on a value between 1 and 9 (Table 6.3, Figure 6.4).

The set of results forms the pairwise comparison matrix of the criteria – in our case, the barriers considered (Table 6.4). Each of the decision makers then carried out all of the pairwise comparisons in order to qualify the relative importance of each barrier to the adoption of knowledge management practices. The average of all the responses was then calculated, so as to arrive at a consensus within the group of 15 decision makers.

Saaty scale	Corresponding language formulations
1	Equal importance
3	Slightly more important
5	Moderately more important
7	Much more important
9	Very much more important

Table 6.3. *Word-based scale used for the weighting of criteria*

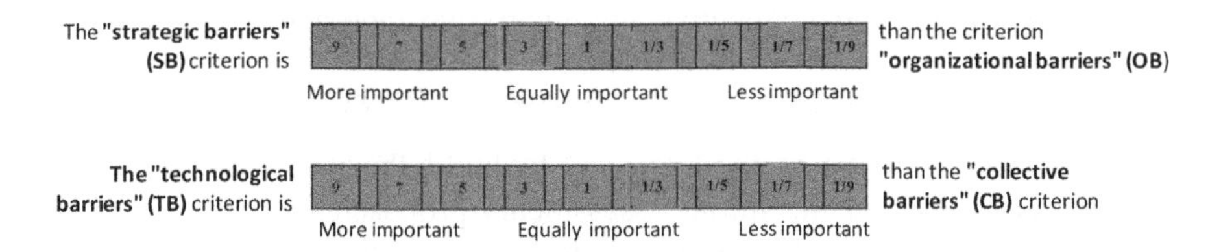

Figure 6.4. *Example of a pairwise comparison. For a color version of this figure, see www.iste.co.uk/enjolras/decisionmaking.zip*

	SB	OB	TB	CB	IB
SB	1	3	9	5	9
OB	1/3	1	5	3	9
TB	1/9	1/5	1	1/3	3
CB	1/5	1/3	3	1	7
IB	1/9	1/9	1/3	1/7	1

Table 6.4. *Pairwise comparison matrix*

This consensual comparison matrix then gave rise to the calculation of the weighting of each barrier to the adoption of knowledge management practices within the supply chain (Table 6.5), following the steps of the AHP method: the normalization of the matrix and then the calculation of the average of each row (see Chapter 2).

SB	OB	TB	CB	IB
51%	26%	6%	14%	3%

Table 6.5. *Weighting of criteria*

The consistency of the results was also checked, as allowed by the AHP method. The related consistency index is 6%, which remains below the acceptable threshold which was set at 10%. Thus, the expert group considered that the most important barrier was the strategic barrier, followed by the organizational aspects. The cultural barrier is also of significant importance, while technological and individual barriers are more negligible. These results reinforce the fact that knowledge management practices within the supply chain must move from an operational and technical realm toward a strategic valuation and a collective and integrative approach.

6.2.2.3. *Normalization of the weighted matrix of assessments*

The next step in the methodology is to build the weighted matrix of assessments. Based on the initial assessments proposed by the experts in section 6.2.2.1, and combining this information with the weight of the criteria constructed in section 6.2.2.2, it is then possible to create a new matrix that includes all of these data.

To this end, the evaluation matrix (Table 6.2) was first normalized, so as to only contain values between 0 and 1. The values were then linearized according to the following affine function:

$$E_{norm,i,j} = \frac{E_{i,j} - min(E_i)}{\max(E_i) - min(E_i)}$$

where:

– j is the alternative considered;

– i is the decision criterion considered;

– $E_{i,j}$ is the evaluation of alternative j for criterion i;

– $min(E_i)$ is the minimum of the set of evaluations of criterion i;

– $max(E_i)$ is the maximum of the set of evaluations of criterion i.

Note that by choosing this normalization method, we are choosing an "ideal" method, as opposed to a distributive normalization, which could also have been made. The distributive normalization is calculated by considering the set of evaluations of alternatives as interrelated. For example, to calculate the distributed normalized evaluation of the alternative j for the criterion i, this evaluation is divided by the sum of all the alternatives for this same criterion i. The distributed normalization therefore allows the calculated normalized score to change in the event that the evaluations of the other alternatives change or if some alternatives are deleted or added. On the other hand, the ideal normalization is calculated by using the alternatives considered as better and worse as a basis. These act as a reference base. Thus, the ideal normalization is independent of the evaluations of all the alternatives, apart from those that act as the maximum and the minimum (see the formula above).

In general, it has been shown that there is no fundamental difference related to the use of one standardization mode or the other (Ishizaka and Nemery 2013). On the other hand, the choice can be argued by the posture of the decision maker. In the case where the latter is sensitive to the strength with which one alternative dominates the others (i.e. the gap existing between the evaluations for a given criterion), then a distributed method would be more appropriate. In the case when the decision maker is more sensitive to how close an alternative is to an ideal value (i.e. how close the evaluation approaches the maximum), then an ideal mode would be best (Schmoldt et al. 2001). In applying TOPSIS, the two modes of normalization can in fact be chosen, but this methodology inherently relies on a comparison of the distance with a desirable maximum and an undesirable minimum, and we made the choice to use the ideal normalization in this case.

Solutions	SB	OB	TB	CB	IB
S1	1	0.75	0	0.5	0.25
S2	0.75	1	0.25	1	0.75
S3	0.75	0.25	0.25	1	0.75
S4	0.5	0.5	0	1	1
S5	0.5	0.5	0.5	0.25	0.25
S6	0.5	0.5	0	0.5	0.5
S7	0	0.75	1	0	0
S8	0.75	0.5	0.25	0	0
S9	0.25	0.5	0.75	0	0
S10	0	0.75	0.25	0.5	0.5

Table 6.6. *Normalized assessment matrix*

Then, each evaluation $E_{norm,i,j}$ was weighted according to its relative importance. The evaluations of solutions according to strategic, organizational, technological, cultural and individual barriers (Table 6.6) were then multiplied by 0.51, 0.26, 0.06, 0.14 and 0.03, respectively (Table 6.5). This allows us to generate the weighted matrix of evaluations (Table 6.7).

Solutions	SB	OB	TB	CB	IB
S1	0.51	0.20	0.00	0.07	0.01
S2	0.38	0.26	0.01	0.14	0.02
S3	0.38	0.07	0.01	0.14	0.02
S4	0.25	0.13	0.00	0.14	0.03
S5	0.25	0.13	0.03	0.03	0.01
S6	0.25	0.13	0.00	0.07	0.02
S7	0.00	0.20	0.06	0.00	0.00
S8	0.38	0.13	0.01	0.00	0.00
S9	0.13	0.13	0.04	0.00	0.00
S10	0.00	0.20	0.01	0.07	0.02

Table 6.7. *Normalized and weighted matrix of assessments*

6.2.2.4. *Construction of the best and worst alternative and calculation of Euclidean distances*

It is then appropriate to apply the computational steps of the TOPSIS method in order to evaluate the distances of each of the solutions with respect to the best and worst alternative.

In the case of our decision problem, the framework of criteria represents the barriers to the adoption of knowledge management practices within the supply chain, and the evaluations were formulated as the ability of a solution to overcome these barriers. Low evaluation values indicate a limited ability to overcome these barriers, while high values reflect a strong ability to address them. Therefore, all criteria are to be maximized. The best alternative is therefore the solution for which the evaluations are maximum for all the criteria, while the worst alternative is the one that is represented by the minimum values of each criterion. Thus, the solutions A+ and A-, presented in Table 6.8, have been defined as the best and worst alternatives for the previous Table 6.7.

	SB	OB	TB	CB	IB
A+	0.51	0.26	0.06	0.14	0.03
A-	0	0	0	0	0

Table 6.8. *Definition of A+ and A-*

The best alternative (A+) therefore represents an ideal dummy solution that would allow the decision maker to act simultaneously on all the barriers to the adoption of knowledge management practices with maximum effectiveness on each of them. Thus, the impacts of solution A+ on the barriers are distributed according to the weight that has been assigned to each criterion by the decision maker. The sum of the evaluations of A+ is equal to 1, representing the optimum value which all values obtained should approach. On the contrary, the worst alternative A- represents a dummy solution that would not help to address any of the identified barriers. Its impact is zero on all criteria, and all evaluations are at 0. Thus, solution A- represents a configuration that the decision maker will seek to avoid as much as possible.

6.2.2.5. *Calculation of Euclidean distances and classification of solutions*

The calculation steps previously discussed (section 6.2.1.2) thus generated the following results (Figure 6.5).

The ranking reflects the capacity of the solutions S1 (*establish positive leadership*), S2 (*encourage teamwork and transparency*), S3 (*strengthen cultural cohesion and cooperation*) and S8 (*build partnerships*) to address the most significant barriers identified. These solutions mainly refer to the importance of the role played by top management, and in particular through its involvement in approaches to knowledge management and the dynamics in place within the company. They also highlight the importance of collaborative aspects both at the level of work teams and with all actors in the supply chain. On the other hand, the solutions S7 (*use digital tools*), S9 (*strengthen the data collection process*) and S10 (*disseminate collaborative practices*) are classified as being a low priority. They mainly concern the more operational aspects of the implementation of knowledge management practices within the supply chain: the means available (tools), the procedures (data collection) and the day-to-day collaborative learning processes. Thus, it would appear that in order to facilitate the adoption of knowledge management practices within its supply chain, company X must first act at a strategic level to gradually move down to the operational dimension of these internal processes. In order to be part of a progressive and sustainable approach, it is first necessary to create favorable conditions (strategic drivers) to consider longer term actions at the operational level.

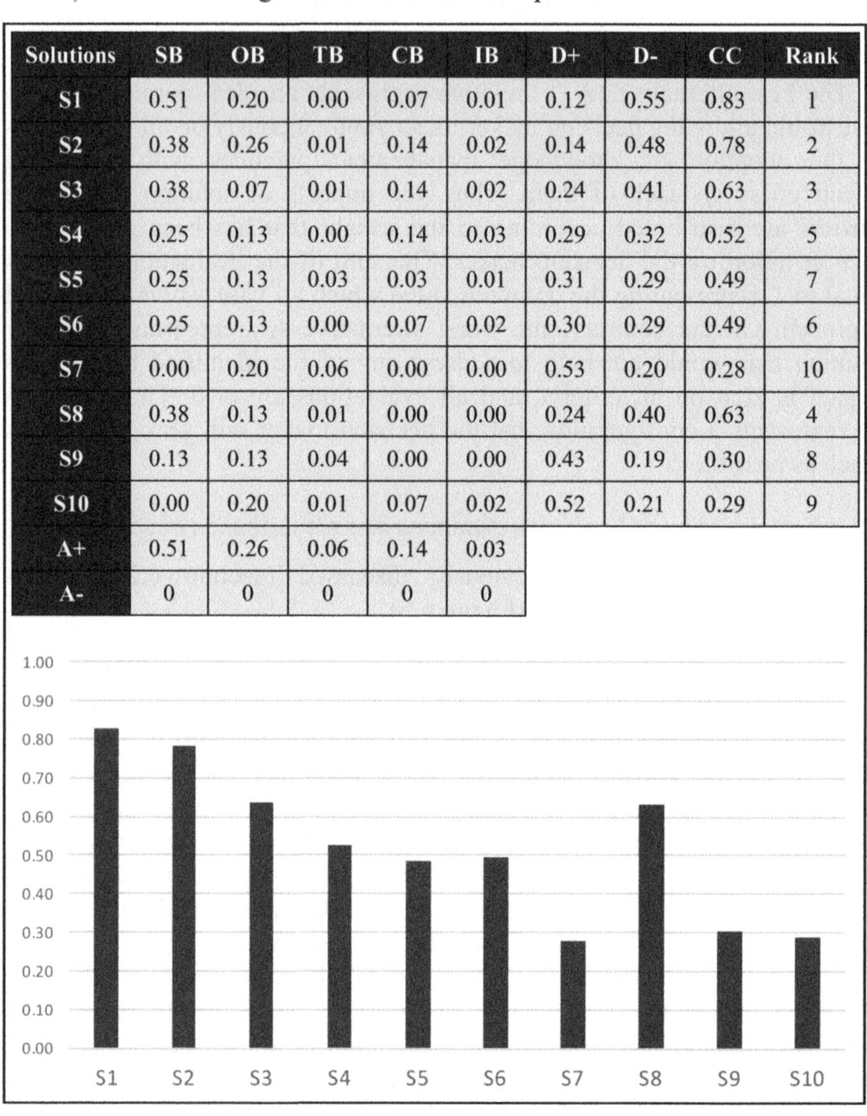

Solutions	SB	OB	TB	CB	IB	D+	D-	CC	Rank
S1	0.51	0.20	0.00	0.07	0.01	0.12	0.55	0.83	1
S2	0.38	0.26	0.01	0.14	0.02	0.14	0.48	0.78	2
S3	0.38	0.07	0.01	0.14	0.02	0.24	0.41	0.63	3
S4	0.25	0.13	0.00	0.14	0.03	0.29	0.32	0.52	5
S5	0.25	0.13	0.03	0.03	0.01	0.31	0.29	0.49	7
S6	0.25	0.13	0.00	0.07	0.02	0.30	0.29	0.49	6
S7	0.00	0.20	0.06	0.00	0.00	0.53	0.20	0.28	10
S8	0.38	0.13	0.01	0.00	0.00	0.24	0.40	0.63	4
S9	0.13	0.13	0.04	0.00	0.00	0.43	0.19	0.30	8
S10	0.00	0.20	0.01	0.07	0.02	0.52	0.21	0.29	9
A+	0.51	0.26	0.06	0.14	0.03				
A-	0	0	0	0	0				

Figure 6.5. *Ranking of solutions*

6.3. To go further

In this section, we will take a more detailed look at some aspects regarding the consideration of the imprecision and subjectivity inherent to each decision-making process. The case study we considered reflects a relatively basic application of the TOPSIS method applied to knowledge management in the supply chain. However, to truly approach the complexity of this decision-making problem, several improvements can be proposed.

6.3.1. *The imprecise nature of human judgment: moving towards "fuzzy logic"*

Within the framework of classical logic, propositions are binary: they are either true or false. However, since human judgment is inherently vague and imprecise, it is sometimes difficult to make an estimate of it through exact and binary numerical values. With this in mind, "fuzzy logic" was devised by Lotfi Zadeh in 1965 to model the imprecision of human knowledge. It has since been applied in many fields such as automation, signal processing, robotics, image processing. Thus, fuzzy logic makes it possible to transform a human linguistic estimate into a numerical value. But this numerical value is then referred to as "fuzzy" because it associates a degree of affiliation with an interval. This degree of affiliation is between 0 (where the value is not affiliated with the interval) and 1 (the value is fully affiliated with the interval, and it is an exact value). Therefore, this makes it possible to represent a degree of confidence with respect to the given numerical value.

In this way, by integrating fuzzy logic into the decision-making problem addressed in this chapter, it would be possible to potentially take into account the subjectivity related to imprecise human judgment at different levels:

– solution evaluations (judgments made by a group of experts, based on a five-level linguistic scale; see section 6.2.2.1);

– pairwise comparisons of criteria to construct relative weights (comparisons based on a nine-level word-based scale, the Saaty scale; see section 6.2.2.2).

Indeed, fuzzy logic is regularly used in conjunction with the TOPSIS methodology to manage the imprecision associated with the issues involved in the decision under consideration, and especially at the stage of evaluating alternatives to be classified (Nădăban et al. 2016). In our case, by proposing that experts use a linguistic judgment scale to evaluate the potential solutions, and by relying on a "fuzzification" approach (application of fuzzy logic TOPSIS), it is then possible to model the imprecision related to these judgments.

One of the operational solutions for applying fuzzy logic is the use of triangular fuzzy numbers (TFNs). A TFN is a three-value set denoted as $\tilde{n} = (n1, n2, n3)$, where "$n1$" represents the lowest probable value, "$n2$" the most likely value and "$n3$" the highest possible value. It is then possible to calculate the degree of belonging of a value x to this TFN by considering a triangular function as follows:

$$\mu(x) = \begin{cases} 0\,,\; x < n1 \\ \dfrac{x - n1}{n2 - n1}, n1 \leq x \leq n2 \\ \dfrac{n3 - x}{n3 - n2}, n2 \leq x \leq n3 \\ 0\,, x > n3 \end{cases}$$

where:

– x is an exact value declared;

– $\tilde{n}$ $(n1, n2, n3)$ is a TFN representing a defined linguistic variable.

Thus, to give a concrete example of the application of "Fuzzy TOPSIS", let us consider the scale used by experts to evaluate the solutions considered to facilitate the adoption of knowledge management practices within the supply chain. By converting what was initially a numerical scale into a word-based scale associated with TFNs, we obtain Table 6.9.

Initial scale	Fuzzy scale	TFN (Patil and Kant 2014)	Associated word variable
1	$\tilde{1}$	(1,1,2)	Very low ability to address the barrier
2	$\tilde{2}$	(1,2,3)	Low ability to address the barrier
3	$\tilde{3}$	(2,3,4)	Moderate ability to address the barrier
4	$\tilde{4}$	(3,4,5)	High ability to address the barrier
5	$\tilde{5}$	(4,5,6)	Very high ability to address the barrier

Table 6.9. *Fuzzification of the solution evaluation scale*

When evaluating solution A, if the expert declares that the solution's ability to address barrier 1 is "low" and assigns it the value of 2, then solution A belongs to the TFN $\tilde{2}$ and depending on the degree of belonging set for the decision problem, it could potentially result in an exact value between 1 and 3.

The definition of this exact value (or "defuzzification") can then be carried out using various computational methods presented in the literature. One of the most common of these is the "α-cuts" for which the decision maker defines a value of μ, a degree of belonging considered acceptable (Bouchon-Meunier and Zadeh 1995).

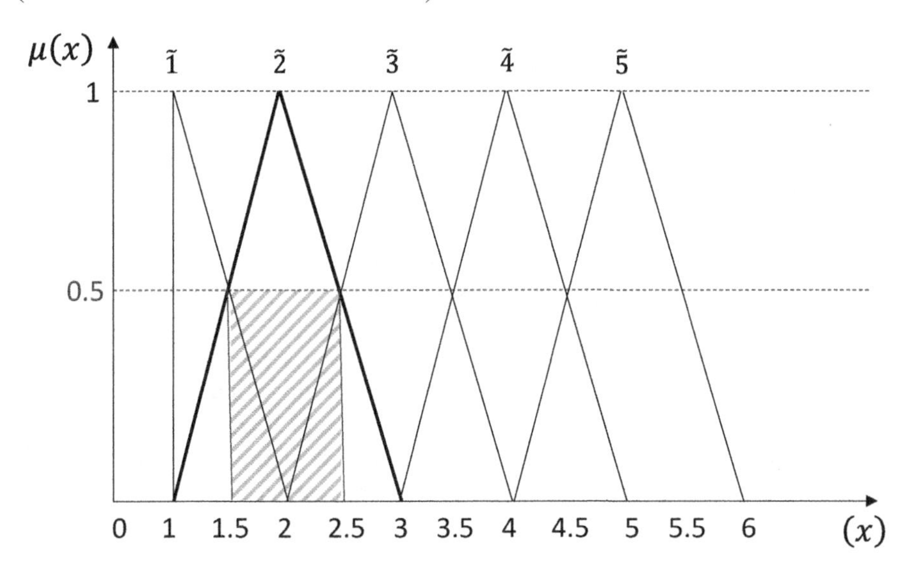

Figure 6.6. *Illustration of TFNs for a degree of membership of 50%*

If the choice is made to consider a minimum degree of belonging $\mu = 0.5$, then for our example, the possible values are reduced to the interval [1.5; 2.5], as indicated by the graphical representation presented in Figure 6.6.

Then, as proposed by Lai and Hwang (1992), a weighted average of the possible values of the interval under consideration is calculated by assigning a weight to the lowest value (pessimistic value), a weight to the most likely value and a weight to the highest value (optimistic value).

The exact value is then obtained using the following formula:

$$x_{exact} = w_{pessimistic} \cdot x_{min} + w_{prob} \cdot x_{prob} + w_{optimistic} \cdot x_{max}$$

where:

– x_{min} is the minimum value of the interval;

– $w_{pessimistic}$ is the weight affected to the pessimistic value;

– x_{prob} is the central value of the interval;

– w_{prob} is the weight affected to the probable value;

– x_{max} is the maximum value of the interval;

– $w_{optimistic}$ is the weight affected to the optimistic value.

Thus, the application of fuzzy logic in this type of decision-making problem adds an additional layer of consideration, taking into account the imprecision and uncertainty involved in the human decision. We have illustrated the application of this fuzzy logic to the solution evaluation stage through a presentation of "Fuzzy TOPSIS".

But this logic can also be applied to all multi-criteria analysis methods. For example, it is quite conceivable to apply this fuzzy logic to the construction of the weightings of the criteria using "fuzzy AHP". Indeed, "fuzzy AHP" is an extension of the traditional method of constructing weights by pairwise comparisons. It allows us to restrict the limits related to the imprecision of the comparison scale (Saaty scale – between 1 and 9).

It is then a matter of improving this comparison scale by replacing its values with fuzzy values to which a degree of affiliation is assigned through the use of TFNs. This also limits the imprecision of the comparisons made by the decision maker in the constitution of the decision model.

6.3.2. *Sensitivity analysis or the proposal of simulation scenarios*

A sensitivity analysis is a study of the influence of the input data of a system on the output data of that system. Many applications of these analyses can be conceived: testing the robustness of a system, quantifying the uncertainty of the system, understanding the relationships between variables or even carrying out simulations and projections.

In the proposed case of application, a sensitivity analysis is particularly useful in testing the changes in the classification of solutions to be implemented depending on the modification of the weights assigned to the barriers. This makes it possible to envision different scenarios where there is variance in the importance assigned to a given barrier to the adoption of knowledge management practices. This involves carrying out simulations in which the strategic context of the company changes.

In this way, various scenarios have been considered (Table 6.10). The first scenario is that obtained by considering the weights defined previously. This is the initial scenario. The following five scenarios, respectively, give the advantage to one type of barrier, which becomes dominant in terms of affected weight. Scenario 2 gives special importance to strategic barriers, scenario 3 emphasizes the importance of organizational barriers, scenario 4 technological barriers and so on. Finally, the last scenario (7) proposes an equal weighting between the criteria.

	Scenario 1	Scenario 2	Scenario 3	Scenario 4	Scenario 5	Scenario 6	Scenario 7
	Initial	SB dominant	OB dominant	TB dominant	CB dominant	IB dominant	Equal weightings
SB	51%	60%	10%	10%	10%	10%	20%
OB	26%	10%	60%	10%	10%	10%	20%
TB	6%	10%	10%	60%	10%	10%	20%
CB	14%	10%	10%	10%	60%	10%	20%
IB	3%	10%	10%	10%	10%	60%	20%

Table 6.10. *Presentation of scenarios for sensitivity analysis*

These changes in the weights assigned to the criteria will then have a direct impact on the classification of solutions to promote the adoption of knowledge management practices within the supply chain (Figure 6.7). Thus, each scenario reflects a certain context for which a different prioritization of alternatives can be identified. If the company assigns great importance to the reduction of strategic barriers (scenario 2), then the solutions to be implemented as a priority are S1 (*establish positive leadership*), then S2 (*encourage teamwork and transparency*) and S3 (*strengthen cultural cohesion and cooperation*). On the other hand, if, for example, the company wishes to prioritize solutions that address technological barriers (scenario 4), then it will have to prioritize the implementation of solutions S7 (*use digital tools*) and S9 (*strengthen the data collection process*). It is interesting to note that solution S2 (*encourage teamwork and transparency*) appears to be the one that is unanimous, with the exception of scenario 4 for which it is not a priority. It appears to provide an overall consensual impact.

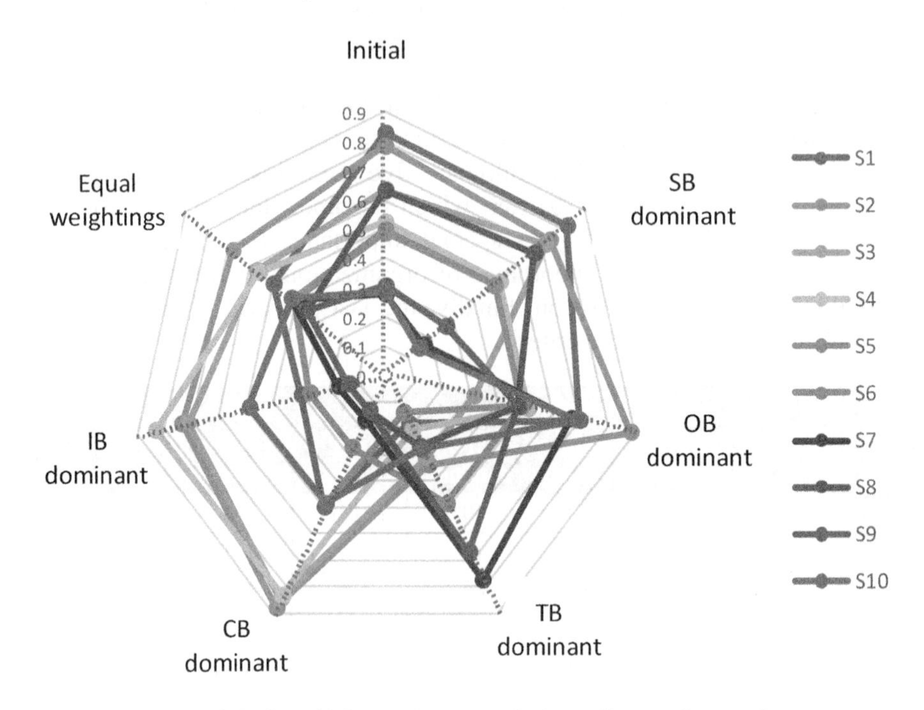

Figure 6.7. *Sensitivity analyses carried out. For a color version of this figure, see www.iste.co.uk/enjolras/decisionmaking.zip*

6.4. The TOPSIS method: instructions for use

6.4.1. *TOPSIS step by step*

To apply the TOPSIS method, this practical guide offers three steps to follow:

– creation of the weighted evaluation matrix;

– identification of the best and worst alternative and calculation of Euclidean distances;

– calculation of the closeness coefficient.

To illustrate this step-by-step application, we will apply the method to a deliberately simplified decision-making problem: the purchase of a new residence. In this sense, we will consider different housing options offered for sale within an urban geographical area. The decision maker is a resident of this geographical area who wants to acquire a real estate property meeting the following criteria:

– a living area of about 70 m^2;

– closeness to the city center;

– a minimum cost;

– maximum energy efficiency performance.

Four properties that were more or less in line with these criteria were listed and evaluated by a local real estate agency (Table 6.11).

Properties	Living area (m^2)	Distance to the city center (km)	Sale price (€)	EPD[1] (A–G)
Apartment	60	0.5	115,000	C
Single-family house	85	6	180,000	A
Duplex	75	2	125,000	E

Table 6.11. *Evaluation of housing units*

6.4.1.1. *Creation of the weighted evaluation matrix*

The decision maker was first asked to provide the importance to be assigned to each criterion. This decision maker indicated that the most important factors were the selling price and the distance from the city center. The other two criteria, the living area of the property and its energy performances, are also important, but to a lesser extent. The weight distribution was therefore established as given in Table 6.12.

1 EPD: energy performance diagnosis. Assessment of the energy performance of a residential unit on a scale ranging from A to G, giving energy consumption in $kWh/m^2/year$. A represents the best energy performance and G the worst.

Criterion	Living area	Distance to the city center	Sale price	EPD
Weighting assigned	20%	30%	30%	20%

Table 6.12. *Weights indicated by the decision maker*

CAUTIONARY NOTE 1.– The definition of the weights of the criteria is a crucial step in each decision-making problem. It must be done with the decision maker and be as objective as possible. To do this, it is sometimes easier to rely on tools and methods rather than statements. That is why the step of formulating the weightings, included in the AHP method, is often considered in the process, thus forming hybrid methods (such as the application case presented in this chapter: AHP/TOPSIS). For the sake of simplicity, for this step-by-step guide, we opted to declare the weightings, but the AHP method could also be applied.

Then, the evaluations proposed by the real estate agency were analyzed, so as to construct the matrix of evaluations reflecting the objective associated with each of the criteria.

CAUTIONARY NOTE 2.– This criteria analysis step allows the decision maker to ensure that each criterion is measurable, that the necessary data are available and that it reflects the elements the decision maker wants to bring to bear on the process. Therefore, each criterion should be constructed as accurately as possible, asking: What is its ideal value (maximization/minimization/target value)? How is it measured (which indicator)? Where is the information found? Is there any redundancy between the criteria constituting the decision problem, and are they sufficiently exhaustive (see Introduction)?

The criteria have therefore been specified as follows (Table 6.13):

– the living area criterion: the evaluation for this criterion should be as close as possible to an area of 70 m^2. The data in Table 6.11 have therefore been modified to reflect a target value to be achieved:

$$E_{area} = |(target\ value - living\ area)|$$

The modified estimates therefore no longer reflect an area as such, but rather the absolute value of the deviation from the target value (ideal area = 70 m^2). Thus, the gap is to be minimized;

– the criterion of distance from the city center has been preserved as it is; it is also to be minimized;

– similarly, the criterion of selling price criterion has been maintained and is to be minimized;

– the EPD criterion is represented on a scale between 1 and 7 in which an evaluation of 1 represents a grade of G according to the EPD standard, while an evaluation of 7 represents a grade of A, and therefore maximum energy efficiency performance according to the DPE standard. This criterion is therefore kept as it is and is to be maximized.

Properties	Living area (m^2)	Distance to the city center (km)	Sale price (€)	EPD (1–7)
Apartment	10	0.5	115,000	5
Single-family house	15	6	180,000	7
Duplex	5	2	125,000	3
Total	30	8.5	420,000	15

Table 6.13. *Modified ratings*

Then, the obtained matrix 6.13 was normalized (Table 6.14). A distributed method was chosen. This involves dividing each term by the sum of the evaluations of the corresponding column:

$$E_{norm,i,j} = \frac{E_{i,j}}{\sum_{j=1}^{N} E_{i,j}}$$

where:

– i is the decision criterion considered;

– j is the alternative considered;

– N is the number of alternatives;

– $E_{i,j}$ is the evaluation of alternative j for criterion i.

Finally, these normalized evaluations were weighted by multiplying them by the weight defined for each criterion, in order to arrive at the normalized and weighted matrix of the evaluations (Table 6.14):

$$E_{W,i,j} = E_{norm,i,j} \cdot w_i$$

where:

– i is the decision criterion considered;

– j is the alternative considered;

– w_i is the weighting of criterion i;

– $E_{norm,i,j}$ is the normalized evaluation of alternative j for criterion i.

CAUTIONARY NOTE 3.– As discussed in the previous section, there are several methods of normalization: ideal or distributive. While the choice of one or the other does not in and of itself change the results obtained in a fundamental way, it is also necessary to question the relevance of one or the other, depending on the context of the problem addressed in the decision (see section 6.2.2.3).

	Normalized assessment matrix				Normalized and weighted assessment matrix			
Properties	**Living area**	**Distance to city center**	**Sale price**	**EPD**	**Living area**	**Distance to city center**	**Sale price**	**EPD**
Apartment	0.33	0.06	0.27	0.33	0.07	0.02	0.08	0.07
Single-family house	0.50	0.71	0.43	0.47	0.10	0.21	0.13	0.09
Duplex	0.17	0.24	0.30	0.20	0.03	0.07	0.09	0.04

Table 6.14. *Normalized and weighted matrix of evaluations*

6.4.1.2. *Identification of the best and worst alternative and calculation of Euclidean distances*

Based on the evaluation matrix, the two dummy alternatives A+ and A- were defined. They represent, respectively, the most desirable (the ideal residence) and the least desirable alternative (the residence to be avoided).

CAUTIONARY NOTE 4.– The definition of the best and worst alternatives A+ and A- directly depends on the targets associated with each criterion.

As the decision problem includes both criteria to be maximized (the EPD) and criteria to be minimized (differences in living area, the distance and the price), each of these non-existent alternatives will be formed from the minimum or maximum of the evaluations of each criterion. Thus, the best alternative A+ represents the property with a living area closest to 70 m^2 (the minimum on the floor area criterion), the closest to the city center (minimum on the distance criterion), the cheapest (minimum on the price criterion) and finally, the most energy efficient (maximum on the EPD criterion). Conversely, the worst alternative A- represents the property with the largest divergence from the ideal living area of 70 m^2 (the maximum on the surface criterion), the furthest from the city center (maximum on the distance criterion), the most expensive (maximum on the price criterion) and finally, the least energy efficient (minimum on the DPE criterion). These dummy alternatives thus serve as a reference point for the calculation of the Euclidean distances D+ and D- (Table 6.15):

$$D_j^{\ +} = \sqrt{\sum_{i=1}^{n}[E_{w,i}(\mathrm{A}+) - E_{w,i,j}]^2}$$

where:

– j is the alternative considered;

– i is number of criteria in the decision problem;

– $E_{w,i}(\mathrm{A}+)$ is the normalized and weighted evaluation value of alternative A+ for criterion i;

– $E_{w,i,j}$ is the normalized and weighted evaluation value of alternative j for criterion i.

Similarly, the Euclidean distance separating each solution from the worst alternative A- is calculated using the following formula:

$$D_j^{\ -} = \sqrt{\sum_{i=1}^{n}[E_{w,i}(\mathrm{A}-) - E_{w,i,j}]^2}$$

where:

– j is the alternative considered;

– i is the decision criterion considered;

– n is the number of criteria in the decision problem;

– $E_{w,i}(\mathrm{A}-)$ is the normalized and weighted evaluation value of alternative A- for criterion i;

– $E_{w,i,j}$ is the normalized and weighted evaluation value of alternative j for criterion i.

Properties	Living area	Distance to city center	Sale price	EPD	D+	D-
Apartment	0.07	0.02	0.08	0.07	0.04	0.20
Single-family house	0.10	0.21	0.13	0.09	0.21	0.05
Duplex	0.03	0.07	0.09	0.04	0.08	0.16
A+	0.03	0.02	0.08	0.09		
A-	0.10	0.21	0.13	0.04		

Table 6.15. *Calculation of Euclidean distances with respect to A+ and A-*

6.4.1.3. *Calculation of the closeness coefficient and interpretation*

Finally, the last step is to calculate an overall score for each property, influenced by a closeness coefficient CC. The highest values of this coefficient are associated with the properties closest to the ideal alternative A+ and farthest from the alternative A-. On the contrary, the lowest values are given to the properties farthest from A+ and closest to A-:

$$CC_j = \frac{D_j^-}{D_j^- + D_j^+}$$

where:

– j is the alternative considered;

– D_j^- is the Euclidean distance separating solution j from the worst alternative A-;

– D_j^+ is the Euclidean distance separating solution j from the best alternative A+.

Properties	D+	D-	CC
Apartment	0.04	0.20	0.83
Single-family house	0.21	0.05	0.20
Duplex	0.08	0.16	0.68

Table 6.16. *Calculation of the closeness coefficient*

In view of the results obtained in Table 6.16, it would appear that the best choice for the decision maker is the apartment, followed by the duplex, and finally the single-family house.

CAUTIONARY NOTE 5.– As with any decision support process, the result is not necessarily the main objective or an absolute truth. The resulting ranking should be discussed with the decision maker and enriched with elements such as sensitivity analyses to test its robustness, as well as the influence of the conditions of the decision problem.

This illustration of the TOPSIS method as applied to a simple case allows us to showcase the fundamental principles of this method, as well as how simple it is to implement. The TOPSIS method thus makes it possible to present a mathematical closeness with respect to both an ideal value desired by the decision maker, as well as an anti-ideal to be avoided. This transposition of the decision maker's preferences into Euclidean distances allows for the creation of a decision support model. Although this example has been deliberately simplified, these steps represent the traditional process of solving a decision-making problem with the TOPSIS method.

6.4.2. *Illustration of related free software*

Although the TOPSIS method is based on relatively simple computational steps, in this section, we will look at some of the software tools used to facilitate its implementation. We will focus more particularly on the freeware program Decision Radar[2], which offers, among others, a fast and free online application of the TOPSIS method.

Let us return to the example of buying a residential property, presented above. The software allows the decision maker first to define up to seven different alternatives. In our case, the three alternatives to be considered are an apartment, a single-family house and a duplex (Figure 6.8).

Next, we will need to define the criteria used in the decision problem. In our case, four criteria are to be taken into account (Figure 6.9): the living area (understood as the divergence from the desired area), the distance from the city center, the sale price and the EPD. We note that this program allows

2 Available at: www.decision-radar.com/.

us to consider criteria that are to be maximized or minimized (the "Negative" column can be checked in the case of a criterion that is to be minimized) but does not allow us to automatically construct criteria with a target value, as was the case for our floor area criterion.

Set Choices

Enter your choice up to 7

Choices	Name
1	Apartment
2	Single-Family House
3	Duplex

Figure 6.8. *Definition of alternatives*

The weightings are considered as input data to be added in the column "Weight". The software does not provide any method for constructing the weightings. Finally, the last column allows us to stipulate whether the criterion used is quantitative or qualitative. In this example, all the criteria are quantitative, that is, they are evaluated by means of an exact numerical value. These numerical values are then entered into the evaluation matrix (Figure 6.10). At this point, we may note a limitation in this program: it only allows integer numeric values to be used. Decimal numbers are not taken into account (this is why the distance from the city center for the apartment is assessed using the value 1 and not 0.5 as in the previous section).

Set Indicators

Enter your indicators up to 7

Choices	Name	Negative	Weight	Qualitative
1	Living area	☑	20	☐
2	Distance to city-center	☑	30	☐
3	Selling price	☑	30	☐
4	EPD	☐	20	☐

Figure 6.9. *Definition of criteria*

Set Decision Matrix

	Living area	Distance to city-center	Selling price	EPD
Apartment	10	1	115000	5
Single-Family House	15	6	180000	7
Duplex	5	2	125000	3

Figure 6.10. *Construction of the evaluation matrix*

If one of the criteria under consideration is qualitative, the method by which this criterion is evaluated changes. Let us imagine that the criterion "distance to the city center" is replaced by the qualitative criterion "interior decoration". Then, the column representing this criterion in the evaluation matrix is presented in the form of an adjustable cursor, to be moved to indicate a given performance (Figure 6.11).

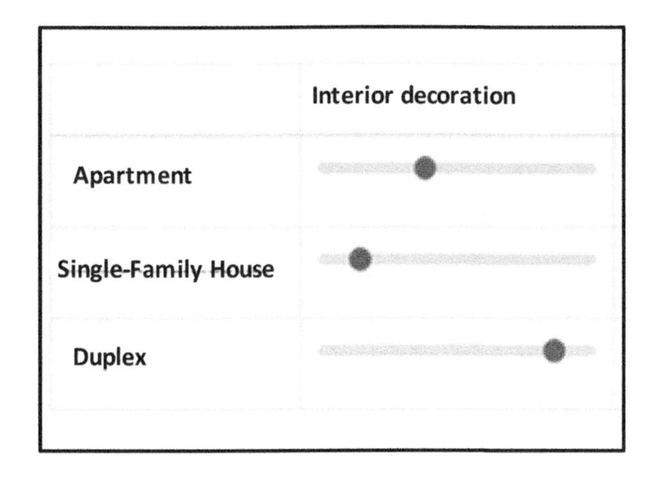

Figure 6.11. *Evaluation of a qualitative criterion*

Finally, the software calculates the results based on the previous three steps. The results are presented in the form of a synthetic window indicating the alternative presenting the best choice for the decision maker, as well as the scores obtained by each of them (Figure 6.12). These scores represent the closeness coefficient CC defined in section 6.2.1.2., shown in this window as Closeness Vector of Each Choice. It is also possible to view the intermediate calculation steps: the normalized evaluation matrix (Normalized Decision

Matrix), the best dummy alternative A+ (Best Answer Vector), the worst dummy alternative A- (Worst Answer Vector), the Euclidean distances D+ and D- (respectively, Choice Distance from Best Vector and Choice Distance from Worst Vector).

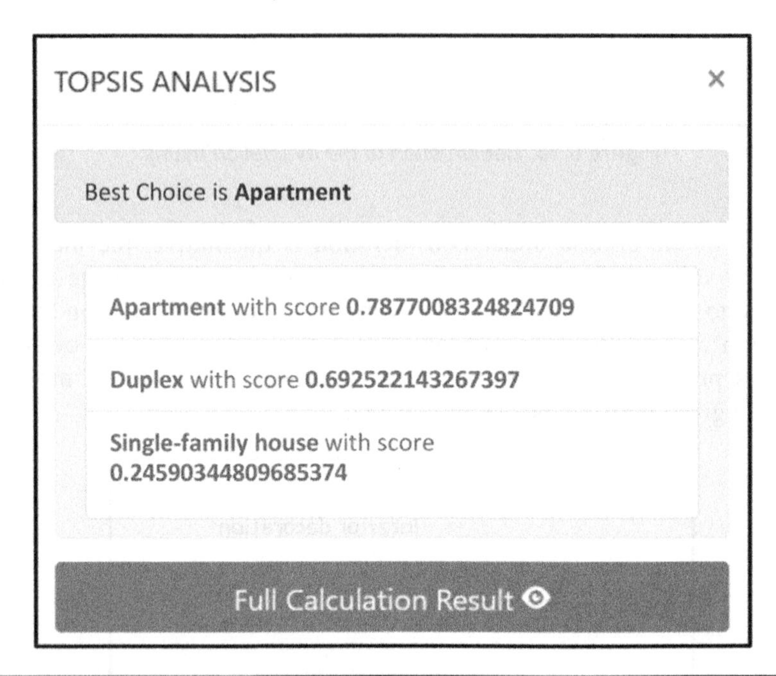

Normalized Decision Matrix is **[[0.10690449676496977, 0.04685212856658182, 0.13940105114820509, 0.10976425998969036], [0.16035674514745463, 0.2811127713994909, 0.2181929496232775, 0.15366996398556648], [0.053452248382484885, 0.09370425713316365, 0.15152288168283162, 0.06585855599381421]]**
Best Answer Vector is **[0.053452248382484885, 0.04685212856658182, 0.13940105114820509, 0.15366996398556648]**
Worst Answer Vector is **[0.16035674514745463, 0.2811127713994909, 0.2181929496232775, 0.06585855599381421]**
Choice Distance from Best Vector is **[0.06917263693481947, 0.2692856911801393, 0.10026417156803168]**
Choice Distance from Worst Vector is **[0.25665359094762397, 0.08781140799175227, 0.22582165663918793]**
Closeness Vector of Each Choice **[0.7877008324824709, 0.24590344809685374, 0.692522143267397]**

Figure 6.12. *Results*

Here, we may note that the results obtained are very similar to those calculated in section 6.4.1. This free software offers a basic application of the TOPSIS method. It does not allow the decision maker to carry out complementary or dynamic analyses, while it does provide a more accessible interface and makes it possible to make the approach more interactive.

TOPSIS AT A GLANCE.– Objective: TOPSIS is a ranking method. Based on the calculation of an aggregated score, it seeks to classify alternatives from the best to the worst, according to the preferences of the decision maker.

Unique features: easy to use and intuitive, it is based on the notion of the distance between the best and worst alternatives.

Limits: it is a partially compensatory method, since it is based on the mathematical principle of the weighted average. Moreover, in cases where all the alternatives considered are bad in view of the preferences of the decision maker, the method then proposes the best possible action among bad options, which may prove unsatisfactory.

6.5. References

Benhamou, P., Bézard, J.-M., Ermine, J.-L. (2002). La gestion des connaissances, un levier de l'innovation. *Revue annuelle des Arts et Métiers*, 293–302.

Bouchon-Meunier, B. and Zadeh, L.A. (1995). *La logique floue et ses applications*. Addison-Wesley, Paris.

Gereffi, G. and Fernandez-Stark, K. (2018). Global value chain analysis: A primer. In *Global Value Chains and Development: Redefining the Contours of 21st Century Capitalism, Development Trajectories in Global Value Chains*, Gereffi, G. (ed.). Cambridge University Press, Cambridge.

Hwang, C.-L. and Yoon, K. (1981). *Multiple Attribute Decision Making: Methods and Applications*. Springer-Verlag, Berlin, Heidelberg.

Ishizaka, A. and Nemery, P. (2013). *Multi-criteria Decision Analysis: Methods and Software*. Wiley-Blackwell, Chichester.

Lai, Y.-J. and Hwang, C.-L. (1992). A new approach to some possibilistic linear programming problems. *Fuzzy Sets Syst.*, 49, 121–133.

Nădăban, S., Dzitac, S., Dzitac, I. (2016). Fuzzy TOPSIS: A general view. *Procedia Computer Science, Promoting Business Analytics and Quantitative Management of Technology: 4th International Conference on Information Technology and Quantitative Management. ITQM*, 91, 823–831.

Patil, S.K. and Kant, R. (2014). A fuzzy AHP-TOPSIS framework for ranking the solutions of Knowledge Management adoption in supply chain to overcome its barriers. *Expert Syst. Appl.*, 41, 679–693.

Saaty, T.L. (1980). *The Analytic Hierarchy Process: Planning, Priority Setting, Resource Allocation*. McGraw-Hill, New York.

Schmoldt, D., Kangas, J., Mendoza, G.A., Pesonen, M. (2001). *The Analytic Hierarchy Process in Natural Resource and Environmental Decision Making*. Springer, Berlin.

Schumacher, A., Erol, S., Sihn, W. (2016). A maturity model for assessing industry 4.0 readiness and maturity of manufacturing enterprises. *Procedia CIRP*, 52, 161–166.

Talbi, C.L. (2018). Le Management des connaissances, levier de l'innovation managériale dans les entreprises apprenantes. *Gestion 200*, 35, 73–101.

Zadeh, L.A. (1965). Fuzzy sets. *Inf. Control*, 8, 338–353.

Conclusion

The great upheavals our society is currently experiencing will inevitably lead us to deeply rethink some of our industrial systems and our overall way of life. Climate change, the scarcity of raw materials and water, the need to decarbonize energy sources or health crises such as COVID-19 are phenomena that are occurring simultaneously and are interconnected. Dealing with these phenomena requires an integrated and systemic vision on our part, as well as the establishment of mechanisms of analysis to deal with this complexity.

Thus, the design and manufacture of new products that are more environmentally friendly requires us to identify new ways to source, manufacture, organize supply chains and identify the needs of users. This implies a design, and more broadly, innovation to be done on many levels.

Indeed, simply taking into account the desired characteristics of a product is no longer enough to ensure it will be a success. The nature of the raw materials, the operational conditions of manufacturing, the environmental impact or the safety conditions of the operators must be included from the first stages of design, and this must be done simultaneously.

In addition, technological developments in the digital field in recent years have significantly increased the capacity to generate and the availability of data and have made the analysis and decision-making process more complex. Thus, in order to take advantage of technologies such as those related to Industry 4.0, decision makers need analytical tools and support for decision-making at all stages of development.

On the other hand, if we consider the innovation process under the definition given by the ISO 56002 (2019) standard, we realize that this process occurs broadly, running through all areas of an organization, and therefore intrinsically consists of multi-criteria, multi-actor and multi-level characteristics. Indeed, the innovation process within an organization can only operate if it is supported by an innovation management system which needs to be put in place. This approaches a level of organization where the decisions to be taken do not depend simply on the expertise of an isolated decision maker, but rather are in a perpetual relationship with all actors potentially involved. In this context, multi-criteria methods appear as powerful tools to objectify and support decision-making within the innovation management system.

From an operational point of view, the application of multi-criteria methods can be summarized in three main stages: (1) definition or modeling of the decision-making problem, (2) choice and application of the preferred multi-criteria method for solving the problem, and (3) interpretation of the results.

Modeling the problem consists of defining the frame of reference for decision-making, determining the choice to be made, the alternatives, the evaluation criteria, as well as their weighting and their evaluation scales. The proper definition of the problem allows the decision maker to better understand the consequences of the decision to be made, considering all the factors involved and the related preferences. This understanding and structuration of the problem is the first result of the application of all multi-criteria methods.

The second stage involves the choice and application of the multi-criteria method in order to prioritize alternatives for solving the problem. This choice of the multi-criteria method to be used depends on the characteristics of the problem, since it is not possible to ensure that one method is better than another, but it is possible to determine which method best suits a particular form of decision making. To choose the method, it is recommended to analyze the purpose for which the decision will be made (the choice of an alternative, classification of alternatives, definition of decision rules, etc.) and the type of evaluation of criteria (qualitative or quantitative). For example, one of the objectives of the decision problem is to define the weighting of the criteria; in this case, AHP and DEMATEL can be used. In the case where the objective is to utilize an existing database to understand how decisions are made, the Rough Sets method is the best option. This choice of the method to be applied represents a major challenge of multi-criteria analysis. Even if all the methods have different computational procedures, once chosen, their application remains accessible and can be supported by different software programs that facilitate their implementation.

Finally, the interpretation phase of the results seeks to highlight the advantages of the preference model that has been constructed and makes it possible to offer the most informed support for the decision as possible. In most methods, the primary result is to produce an overall score for each alternative, representing the degree to which each of them is preferable to the decision maker. If the objective is to determine the preferred alternative for the decision maker, then it is sufficient to rank these overall scores. But the application of the methods allows a much more in-depth analysis, which goes beyond merely pointing to the highest-scoring alternative to choose. Above all, it is about offering support for the decision, and not of an outside process reaching a decision in place of the actors involved. Thus, understanding what makes an alternative preferable, identifying the contributions of each criterion in the decision and proposing scenarios that can simulate specific contexts (sensitivity analysis) are the ways to enrich decision-making alongside decision makers in the most clear and objective way possible.

We do not pretend that multi-criteria analysis is a silver bullet. Nevertheless, its use, as well as that of other methods for formalizing and representing the complexity of different systems, is now proving to be a necessity. Making compromises between various aspects, considering various human preferences, or even proposing scenarios to bring the stakeholders of a project into dialogue are all actions that are facilitated by the use of multi-criteria approaches.

This book highlights how multi-criteria analysis can contribute to the search for compromises when it comes to innovating in industrial sectors. The authors sought to convey, in the most educational way possible, how a decision maker can choose and apply the different existing methodologies, depending on the context of application, the type of decision to be made or the quantity and nature of information available.

To this end, the authors illustrated different decision-making issues and highlighted how multi-criteria methods can support them.

Chapter 1 addresses the upstream phases of innovation through the selection of ideas from a creativity workshop. The PROMETHEE method, applied to this issue, makes it possible to highlight the importance of taking into account the context in this decision by simulating different scenarios for which the decision criteria appear to be more or less critical. This notion of sensitivity thus makes it possible to represent the preferences of a decision maker in as fine a way as possible.

Chapter 2 deals with the formulation of products in the field of the chemical industry. Through a combination of AHP and DEMATEL, this chapter proposes a method for constructing the preferences of a decision maker, resulting in the creation of weightings representing the degree of importance of each established decision criterion. By combining these weightings with the contributions of the DEMATEL method, it is then possible to go further with the modeling of the decision problem, by integrating the inter-criteria relationships as a way to gain a closer understanding of the reality and to objectify the context of this decision-making process.

Chapter 3 questions strategic decision-making and the uncertainty of human judgment through an application case related to the marketing of cosmetic products. By using fuzzy logic and the Rough Sets method, this chapter shows how to build a preference model by including inconsistencies or particularities specific to the decision maker. This preference model is then converted into the form of conditional rules representing the best combinations of marketing resources. It is thus easily exploitable because it is understandable and transparent for the decision maker.

Chapter 4 deals with the topic of project portfolio management of a company and proposes an application case in the field of the oil industry. Using the MAUT method, this chapter highlights the importance of the notion of utility as a driver to best represent the preferences of a decision maker and explores the interest in taking into account the potential synergies existing within the decision criteria reference set.

Chapter 5 focuses on a study of the recruitment process, as a key issue for skills management within an organization. To accomplish this, the ELECTRE method sheds light on how to compare potential candidates, approaching this process through the establishment of dominance relationships between different alternatives.

Finally, Chapter 6 seeks to study the adoption of effective knowledge management practices within the supply chain. Using the case of a manufacturing company, this chapter combines the application of the TOPSIS method, fuzzy logic and sensitivity analysis to approach the decision through both the notion of distance from an ideal desired by the decision maker, as well as an anti-ideal to avoid.

Of course, the field of decision engineering, and more precisely multi-criteria analysis, is not limited to the six methods presented here. The authors have made a deliberate choice to present the methods considered as being among the most significant and most frequently used in the scientific literature and the industrial world. However, for the sake of brevity, other equally important methods have not been covered here, including the Ordered Weighted Average (OWA) method, devised by Ronald Yager. Another perspective concerns the extensions of the methods presented here using fuzzy logic (Fuzzy-AHP, for example) to deal with inaccuracies, or in particular, hybrid methods that seek to combine and take advantage of the strengths of the various different methods.

Moreover, another widespread application of multi-criteria analysis methods concerns the exploration of solution spaces in optimization problems. In a multi-criteria problem where the desired outcome is, in particular, to simultaneously optimize several contradictory objectives, this involves finding all the points within the search space, such that there is no other better point on all the criteria simultaneously. If the first phase of the multi-criteria optimization gives too broad a result (the Pareto front), and especially in systems including nonlinearities, the use of a decision support system will make it possible to identify recommendations by modeling the preferences of decision makers. Indeed, in these types of situations, the decision taken should not be seen as optimal, in the mathematical sense of the term, but rather to be a “satisfactory” solution that takes into account a maximum of information (criteria) that bring the problem closer to reality.

Finally, in most problems, the resulting actions or measures are prepared through a long and complex process involving many participants. The innovation process does not come down to one single decision or decision maker, but rather a series of strategies and compromises between various points of view, or between groups that do not share the same solution. In this case, the use of multi-criteria analysis methods in a collective way (Group Decision Making) proves to be very useful to model the preferences of stakeholders in a systematic way, and thus promote dialogue between actors, and therefore to facilitate consensus.

This book is the result of the authors' experiences from various innovation and research projects, but mainly from their course modules, which have been enriched by the exchanges and feedback of their students in recent years at their respective universities. The authors hope that this book can be useful not only for teachers but also for professionals and students seeking to discover and apply the methods of multi-criteria analysis in their projects and activities. The authors now leave it to the inquisitive readers of this book to dig further into this world of multi-criteria methods, one that never stops developing.

List of Authors

Mauricio CAMARGO
ERPI
ENSGSI
Université de Lorraine
Nancy
France

Manon ENJOLRAS
ERPI
ENSGSI
Université de Lorraine
Nancy
France

Christian FONTEIX
ERPI
ENSGSI
Université de Lorraine
Nancy
France

Daniel GALVEZ
Department of Industrial Engineering
University of Santiago de Chile
Chile

Index

E, F

G, H, I

K, M

N, O, P

R

S, T

U, W

Printed and bound by CPI Group (UK) Ltd, Croydon, CR0 4YY

17/08/2023

08101733-0001